How to Restore the
MODEL A FORD

ECHO POINT BOOKS & MEDIA, LLC

How to Restore the Model A Ford was originally published in 1961, making the prices listed in this book no longer applicable. However, one can get a sense of prices in 2013 dollars by multiplying any price you see by 8 (7.81 according to the Bureau of Labor statistics). After 2013, you can find out the multiplier rate by going to: http://www.bls.gov/data/inflation_calculator.htm.

Copyright © 1961 Clymer Publications

Published by Echo Point Books & Media
www.EchoPointBooks.com

ISBN: 978-1-62654-941-8

Cover image by Richard B. Hamilton

Cover design by Adrienne Nunez,
Echo Point Books & Media

ANNOUNCEMENT

As original publishers of books about Ford cars we have published a number of very popular books dealing with the Model A Ford, its history and statistics. Public demand has caused us to issue this volume specifically for the person who is contemplating or proceeding with a restoration project. It is, as far as we know, the most complete book on the Model A from that standpoint. We have included all the dimensions, technical data and operational hints necessary to accompany a restoration from the ground up. For that reason, we are sure that collectors of automobile facts and figures as well as admirers and owners of Model A's will also welcome the book.

In providing this mass of information, much of it from rare Dealer Service Bulletins and other publications from the 1928-1931 era, we are making no attempt to instruct the owner or enthusiast in the use of tools or methods of going about the rebuilding. Inasmuch as each individual has his own ideas about such an enjoyable pastime, we feel that by supplying him with all the available information on the car itself and a general guide to the desired end, we will have best served the purchasers of this book.

The initial chapter on restoration of the A is reprinted from HENRY'S FABULOUS MODEL A, by Leslie R. Henry, a Clymer publication ($4.00) which should be in the hands of every enthusiast if for no other reason than it contains a history of the development of the A and its introduction to a waiting world. Many other items of a general nature not carried in this book will also be found in The FABULOUS A.

For a complete photographed record of every model and body style from the 1928 Standard Phaeton to the 1931 Closed Cab Pickup Truck see Clymer's MODEL A ALBUM, ($3.00) 142 pages of photos, historical accessory ads, commercial vehicle information and nostalgia. Hundreds of factory photographs provide absolutely authentic details of exterior appointments and trim.

I would like to acknowledge the generous contributions of material from the Ford Research and Information Department, The Henry Ford Museum and the Public Relations Department, Ford Division, Ford Motor Co., and hope that the information contained between these covers will help keep alive the present interest and affection for a remarkable automobile.

April, 1974   *Floyd Clymer*

WHAT PRICE AUTHENTICITY?

by Ford Authority Leslie R. Henry

MODEL A's present prestige and capability of inspiring pride of ownership can equal, if not surpass, that of any modern Detroit product—but only when the MODEL A is meticulously and *authentically* restored to "like new" condition. Anything less may be a nice automobile but it *cannot* be MODEL A!

Now, if you think you want to restore a MODEL A, first get one.* They can still be found today as wrecks in junkyards, as relics in old garages or barns, as grossly altered "old used cars" in daily use, or perhaps already restored in some degree. Prices will range from $15 for a junker to $1,600 for an *authentic* restoration.

When you find a MODEL A to restore, don't be discouraged by the hopeless appearance of a dismembered junker, nor fooled by the flashy appearance and paint of an unauthentically restored car. The latter will cost more to buy and has more work for you to *undo!*

Anybody can do *some* MODEL A restoration work himself; many amateurs have done it *all!* Certainly, anybody can do the first step himself—take the car *completely* apart. There is no half way—there is no easier way. The car *must* come apart so that each piece can be properly cleaned, inspected, repaired or replaced, refurbished and repainted like new, and then reassembled into a complete MODEL A.

Whether you are a beginner or not, it is always best to be systematic in dismantling a car for restoration. Keep related parts together in trays or boxes provided in advance; identify unfamiliar parts with tags or marks, and make brief sketches of their assembly, if necessary, so you can put them all back together properly later.

Usually it doesn't pay to rush through a restoration job—or any part of it. If you are temporarily stopped for the want of some part or specialized job you can't do, then set that work aside and take up another phase of the work. It is best to have several such jobs going at the same time so that, even if you have set yourself a deadline for completion, the overall restoration work can progress unhurried.

In every restoration project you will come across at least one problem that will seem insoluble and which may tend to discourage you to the point of giving up—don't! You quite probably can solve the problem with a little time and thought, or you will find someone who can solve it for you. In either case, the experience of having won out will give you immense personal satisfaction—like playing a hole-in-one on the golf course, or bowling a perfect game!

If you, personally, haven't the skill or equipment to weld, or to machine a part, or to paint the body, or to upholster, or to overhaul the engine and transmission—then get help from someone who can. Take such jobs to the local shops; there are many other restoration jobs you *can* do.

Remember always that *authenticity* is imperative—whether in a major item such as using only lacquer on the body and fenders of MODEL A, or whether in a minor item such as mounting the wheel on the spare rack with the valve stem at the top and with the FORD script in a horizontal, easily readable position! Trivial? Perhaps; but such were the details of MODEL A on the assembly line, and such are the details which must be preserved if we want a *real* MODEL A today. It is the attention paid to the trivial today which distinguishes a restoration job as authentic.

While higher quality restoration work is always encouraged, we must caution against "over-restoration." Examples of this may be seen on cars restored by the over-zealous in the form of chromium plated wheel lug nuts; filled, rubbed, and polished chassis and axles; fancy, plastic upholstery; striped engine hood louvres; and chromium plated engine parts. Any over-restoration is as bad as not enough—neither is authentic nor like the original MODEL A.

The accounts of the personal experiences of some of the AACA and MARC members who have recently restored MODEL A Fords are both interesting and illuminating:

Russell J. Gerrits, of Chicago, was brave enough to start with a pile of junk (SEE FIG. 34) "in a quite incomplete state, but in good, solid condition. At first the car was not restored to perfect condition; just given a complete but easy going-over in 1954. We had a very good time with the car during the summer, then it was gone over again for the MARC Meet in Dearborn in 1955, where it won first place. But I was certainly not entirely satisfied with the restoration, so in 1957 the car was again completely torn down and restored *to original* factory specifications (SEE FIG. 35). The chassis and motor were built up with as many *new* parts as possible purchased from Ford dealers or from club members dealing in parts. The car won again in the MARC and finally became the AACA National First Prize winner at Lake Forest in 1958."

Harold G. Fulmer of Allentown, Penna., is a recent AACA member who, as a complete amateur, successfully undertook restoration of the "relic" type of MODEL A. This was a project which included his two teen-age sons and about which he gives more detail: "Finally the boys came home from one of their many searches with the news that they had found a MODEL A Station Wagon. I went with them to look at it and all I could see was a rotted and rusted relic (SEE FIG. 36). It had been standing in a wooded area for twelve years, it had rifle shots through the hood, and it had a cracked cylinder head . . . so we bought ourselves a lot of work . . . for $15!"

The farmer who sold it couldn't figure out what they wanted with it and, after they had towed it home on borrowed wheels and with stops every block or so to pick up parts falling off the body, the neighbors wondered too. "Do you intend to make something out of *that?*" they asked.

Harold considered that a challenge, so he started: "I hosed it off and then took it all apart. Incidentally, we haven't a garage to work in—only a cellar with a 30-inch door. In the following weeks we bought a junk sedan as a parts car. *All* metal parts selected for the Station Wagon project were buffed down to the bare metal, washed in gasoline and taken into the cellar shop. Then each piece was carefully inspected, primed and painted, then made

FIGURE 34. YESTERDAY'S PILE OF JUNK.

Mrs. Gerrits ruefully views the remains of a 1931 Ford De Luxe Roadster in a Wisconsin junkyard and wonders if her husband really should have saved it from the doom of an open hearth furnace.

Russell J. Gerrits, of Chicago, thought so in 1954 and devoted several years of work to its restoration. At first made only roadworthy, the car did not satisfy Russell until he had completely rebuilt it—twice! The finished car now an AACA National First Prize winner, appears in Figure 35.

FIGURE 35. TODAY'S PRIDE AND JOY.

1931 Ford De Luxe Roadster 40-B, owned and restored by Russell J. Gerrits, Chicago, Ill.

No pile of junk this, but a beautifully and authentically restored MODEL A No. 4316729. Twice a first place winner in MARC meets at Ford's Greenfield Village in 1955 and 1957, it was re-restored in 1958 in time to be judged a National First Prize winner by AACA at the Lake Forest (Ill.) Meet.

FIGURE 38. 1928 FORD — CUSTOM BUILT STATION WAGON.

Owned and restored by Harold G. Fulmer, Allentown, Pa.
(Engine No. A210759)

Originally one of the 1928 Ford Commercial Chassis fitted outside the Ford plant with a "custom built" station wagon body, this car was excellently restored, piece by piece, from the rotted relic shown in Figure 36. All work except electroplating and spray painting was done by the owner and his two sons, all of whom were complete amateurs.

FIGURE 36. IT CAN BE RESTORED!

Harold G. Fulmer and his two sons found this discouraging looking relic on a farm near their home in Allentown, Pa. Although new to the hobby and complete amateurs, they restored this 1928 Ford Station Wagon to prize-winning condition at a cost of only $565.98—and plenty of their own labor!

up into sub-assemblies that would pass through the cellar door—front axle, rear axle, engine, frame, cowl, steering, etc.

"The body was rebuilt with the original type hard maple; the old pieces were first laid out on the floor to use as patterns. Fortunately, some of the pieces missing on one side happened to be on the other so we could duplicate all parts. We then set the frame on the axles and then used "C" clamps and a few bolts to assemble the wooden body as a temporary unit to make sure everything fit properly.

"A friend gave me an old treadle sewing machine which we used to sew up exact duplicates of the leatherette seat upholstering and the side curtains. These were made of convertible top material with plexiglas windows. We had to get an upholsterer to sew in the plexiglas because it was too stiff for our old sewing machine.

"Then one Saturday, the two boys and I took the car apart again and carried it outside; the rebuilt engine we skidded up the steps on the boys' sled (SEE FIG. 37). In another week we had the MODEL A reassembled and ready for the black and tan paint. So far, the restored car has won prizes in the AACA Lehigh Valley Region and at other local Meets."

John Mearkle of Springfield, Penna., gives an account of his restoration of a 1931 Standard Roadster which became an AACA National First Prize winner at Pottstown in 1957 (SEE FIG. 39):

"I purchased the car on New Year's Eve, 1950 for seventy-five dollars. The car ran, and that is about all. The body was pretty well rusted out, badly dented and cracked. The rumble seat interior was in an advanced state of decay; the roof was a few tattered shreds of canvas over rotten bows; and the windshield frame was a rusted shell with no glass. The entire mess, as I found out later, was covered with five coats of house paint which actually served to hold some parts together!

"I did the restoration of the body myself, except for the roof, side curtains, and top bows. I made the windshield frame of ⅞" round brass stock, bent it to the shape of the cowl, and channeled it to take the glass and the

COST OF RESTORING A 1928 FORD STATION WAGON BY HAROLD FULMER AND SONS, LEHIGH VALLEY REGION, AACA.

Description	Amount
Price of Station Wagon, November 1, 1957.	$ 15.00
Price of 1928 Ford Sedan parts car.	50.00
Maple wood to remake body.	76.72
New parts: muffler, timing gear, gaskets, windshield wiper, plugs, ignition parts, fan belt, water hose, etc.	47.49
Safety Plate glass windshield.	24.28
Black roof topping.	14.25
Curtain material, canvas, plexiglas, fasteners, etc.	64.12
Upholstering material, tacks, needles, thread.	33.35
Tires—3 new, 2 used.	64.29
*Chrome plating (Original was nickel—AUTHOR'S NOTE)	45.05
Rubber floor mats.	5.78
Paint brushes, nails, glue, varnish, paint remover, nails, screws, nuts, bolts, etc.	22.55
Running board material, counter top moulding, cement.	7.75
Battery and cables.	8.90
*Paint job on body.	89.00
Motometer and steering wheel.	3.00
Chrome plating on motometer, light switch, Ford emblem. (Original was nickel—AUTHOR'S NOTE)	6.50
Total expense	$578.03
Parts car sold for scrap (except motor, transmission and rear axle assembly).	− 12.05
Net cost of restoration.	$565.98

Note: The 1928 Station Wagon was not built by the Ford factory, the body was constructed on the chassis by an outside coachbuilder, so there is no factory price noted. However the original cost of the car in 1928 was probably at least $100 more than the cost of restoration thirty years later.

FIGURE 39. AN AUTHENTIC RESTORATION.
Owned and restored by John A. Mearkle, Springfield, Pa. (Engine No. A4181349)

'T' rubber. Some of the small items, including one set of wing brackets, I made from solid brass stock by cutting, filing, and tapping where necessary. I had all of these items plus the bumpers and other bright work chromium plated.

The channeling supporting the body was so badly rusted through that I had to have new pieces made and welded onto the remaining solid body channel up near the front seat. The rumble seat floor and rear fender wells I made of sixteen gauge galvanized iron riveted into the original solid metal that remained after all the rusted portions were cut away.

"All dents were hammered out of the body, and all cracks were welded and reinforced from the inside. I filled in with solder all the places where I had welded and where I had pieced the body with galvanized iron. Then the entire body was filed and sanded as smooth as possible, and a coat of lacquer primer applied. The remaining small dents and ripples were filled in with lacquer putty and sanded again. The body and fenders were then sprayed with 14 coats of Ditzler lacquer—body color is Colony blue; fenders are jet black.

"The wheels were sandblasted, the spokes straightened, then painted with duPont Miami cream enamel. The frame, axles and under-portions of the body were painted with primer and two coats of black enamel. The engine, transmission, and all the running gear had been completely rebuilt using mostly new parts. I rebuilt the seat and sewed all the upholstery by hand—it took about four months for this, in my spare time.

"Total cost of this restoration (not including my labor) was about $1,800. I think this figure is rather high for a MODEL A, but I didn't pick out a car with reasonable restoring possibilities. Since restoration, I have driven the car 14,000 miles with no major repairs and I consider my restoration costs as money well spent!"

Of course the ultimate of all restoration work is to provide a desirable car to *drive* in addition to reactivating something of American automotive history. The restorer often refers to his efforts as "work" when actually it isn't all all—work is what you are doing when you would rather be doing something else! Just try to get an antique car restorer to do something else when he is in the midst of a restoration job and you'll find out whether or not he is working.

Hardly a collector will not admit that restoration "work" is as much a part of the fun as driving the finished car. The hobbiest who restores his first antique car —MODEL A or other—usually restores another; and then perhaps "just one more!" Nothing gives as much satisfaction as a restoration job well done, and the pride of ownership is enhanced by a pride of personal attainment and workmanship.

Most so-called "masculine" hobbies are strictly that, but MODEL A restoration is often a hobby for the whole family (SEE FIG. 33). The children can usually wield brushes with cleaning solvent on the mechanical parts, or with paint remover on the body, to good effect. And your wife might even take an interest in the trim and upholstering—particularly when she realizes that this way she at least has you home, albeit in the garage!

Of course, the finished car immediately becomes a vehicle of pleasure for the whole family—antique automobiling, particularly with others (SEE FIGS. 25, 41, 71 AND 84) is a new and thoroughly enjoyable family hobby and social activity.

For fullest enjoyment of the hobby you should join one of the many organizations devoted to antique automobiles in general and/or MODEL A Fords in particular. Through such affiliation you can receive help with your restoration work, information in the club magazine, and enjoyable association with kindred souls all enjoying this family hobby at meets, runs, picnics, back-country runs, cross-country tours, and regular social meetings.

It is impossible to list here all the numerous clubs catering to MODEL A enthusiasts, but we can, with propriety and no intent to slight any, list alphabetically those several clubs which are large enough to operate on a national basis with local Regional groups and which are large enough to publish regularly a recognized magazine or bulletin:

Antique Automobile Club of America
West Derry Road, Hershey, Penna. 17033

Horseless Carriage Club of America
9031 East Florence Ave., Downey, Calif. 90240

Model A Ford Club of America
Box 8267, Long Beach, California

Model A Restorers Club
Box 1930A, Dearborn, Michigan

Veteran Motor Car Club of America
15 Newton St., Brookline, Mass. 02146

FIGURE 33. 1929 STATION WAGON — "LIKE HAVING AN OLD FRIEND BACK!"

Owned and restored by Howard G. Henry, North East, Md.

Mr. Henry purchased the car from its second owner for $200 in 1958 and spent an additional $392.31, plus family labor, in its restoration. The car was completely dismantled, then rebuilt like new. With the exception of only three pieces of new maple, the entire body is original. In spite of the hours spent in cleaning, scraping, rubbing, and painting, the Henrys consider the project both rewarding and educational for the whole family.

Photo courtesy The Henry Ford Museum.

FIGURE 25. MODEL A HOMECOMING!

Each year the Henry Ford Museum plays host to MODEL A owners in the Model A Restorers' Club. Here are more than 100 of these fabulous cars assembled at a MARC meet near the "Ann Arbor House" in the Greenfield Village, Dearborn, Michigan. (See also Fig. 84).

Photo courtesy The Atlantic Refining Company.

FIGURE 41. CARS LINING UP FOR GREATEST SHOW ON EARTH.

The largest gathering of antique cars in the world is found annually in the Hershey, Pa. stadium as guests of the Hershey Chocolate Company. Now MODEL A Fords frequently number nearly 60 of the more than 700 cars of all ages and makes at the annual Fall Meet of the Antique Automobile Club of America. (See also Fig. 71.)

Photo by Kenneth Stauffer, AACA Photographer.

FIGURE 71. HERSHEY — THE MECCA OF ANCIENT AUTOS.

MODEL A Fords now figure prominently in national "meets" of the AACA such as this huge gathering for two days of fun, picnicking, and contests in the stadium at Hershey, Pa., in 1958.

Of the more than 700 antique cars here, 60 are the MODEL A Fords assembled in the center foreground, and 100 are the Model T Fords surrounding the track.

Photo courtesy The Henry Ford Museum.

FIGURE 84. FUN WITH FORDS.

Field events and judging for quality and authenticity of restoration are part of the program of this Model A Restorer's Club meet held near the "Secretary House" in the Greenfield Village, Dearborn, Michigan. In the foreground is the early 1928 Ford roadster pickup truck belonging to the Henry Ford Museum, host to the MARC for this annual meet. (See also Fig. 25.)

DO IT YOURSELF? *

On tours, meets, and in clubrooms of the automobilist, one is apt to hear the boast, "And I restored it all myself!" This discourse of pride is generally followed by words of resentment toward those who have their restoration done by perhaps more skilled and experienced mechanics, craftsmen, and painters. With the high standard set by the national clubs, this type of boast is less often heard than, say, five years ago; and today it rings with untruth.

When the hobby first started, the standards of authentic restoration were naturally in a more or less fluid state. The experience of twenty years of devoted research, painstaking labor, and considerable time and expense has resulted in cars as perfect as the day they first saw the light of the local showroom. Perhaps once in the early days of antique automobiling, a person could do all of his own work.

Enhanced by obtaining a running car needing little mechanical work, the restorer generally totalled a general cleaning job, a few new gaskets, tires, and a paint job. This is not to say that in the early days of the hobby, restorations did not equal or excell those of today. This would be unfair to the purists who have staunchly, through perfect, authentic restorations since the beginnings of the hobby, lifted the sights of everyone to its now present high level of perfection. But in those early days, there were relatively few purists and a fellow could restore cars that shone as bright beacons of things to come.

Today, it can be honestly said that it is a rare and unique individual who can restore the average antique automobile by himself. For one, cars located today by and large are in far worse condition than those located in years past. In fact, many a collector is returning to once scornfully rejected, rusted, gutted, half missing cars and is beginning to restore them.

Cars located that are incomplete in body, parts, and other necessaries, because of today's high standard of perfection, require that the one-man restorer be all of the following: machinist, molder, electroplater, wheelright, upholsterer, top maker, painter, etc. etc. ad infinitum, the many specialized crafts that in short reproduce the work of the original factory that made the car. There are a great many hobbyists when given enough time, enough contacts and friends in the various

*Reprinted from BUY AN ANTIQUE CAR, ($3.00) by Scott and Margaret Bailey, Published by Floyd Clymer.

James Howell of Englewood, Colorado is one of those early automobile men who classify both as a master mechanic and a body builder. He began as a blacksmith, turned wagon builder and with the coming of the automobile became an automobile mechanic and body man. With an extensive background of experience and training, Howell is one of the better known professional restorers in the West. Mr. Howell started restoring antique automobiles in the late thirties (photo L. A. Lucas)

shops, foundries, etc., could do all of this work. Whether in each instance the work would equal the standard of the professional craftsman is not possible to say. The average restorer does do a great amount of his own work. But it is virtually impossible to do it all. On the other hand, a great many hobbyists merely do the cleaning up, the paint removing, the wood scraping, and basic long tedious tearing down and cleaning leaving the rest to be farmed out to the various shops in the community.

There is another proportion of hobbyists who transport the car directly from its long spent retirement to a professional, part- or full-time restorer who does the work complete from the ground up.

Every club usually has one or several men who are craftsmen by hobby or by vocation. To these men, fellow club members resort for assistance. One member may be a machinist, another a mechanic working daily at his trade or

Bemis of Brattleboro, Vermont, noted restorer puts the finishing touches on the author's 1905 Model F Ford body

another a painter or an upholsterer. Collectively, they with the other craftsmen of the community contribute anywhere from 60 to 90 percent of the work needed to make the car a potential prizewinner on the fields of club competition.

In answer to the question of whether you should do it yourself, and the answer does not necessarily depend on whether you have sufficient money to have it done, but whether or not you have the time and the ability. Of the latter, ability, we will consider first. It is surprising to learn of doctors, lawyers, teachers, bookkeepers, clerks who never held a spray gun, who have suddenly blossomed out as one of the best automobile painters or masters in one of the other crafts.

It's even more revolutionary to see a beautifully diamond quilted and tufted upholstery touring car done by a man and his wife after being told that that type of upholstering was

Between a 1900 Mobile Steamer and a Model R Ford, Jim Howell takes a brass lamp from his stock. Collectors owe a debt of gratitude to men like Jim Howell and other professional restorers who gained their experience in the early days of automobiling. Their reviving of skills and knowledge would otherwise be lost to our generation (photo L. A. Lucas)

impossible to do today, or if the local shop would undertake the job, the price would be fantastic. So, ability depends not upon experience, but upon the desire to do something that one believes he wants to do and has the time to do.

Today, many a busy hobbyist has his car completely done at the hands of a professional restorer or another club member who has experience and time. The respect of his fellow associates is not lessened by the fact that he never cleaned a single part, scraped a single spoke, nor in any other way contributed to the restoration save to signing the check. His contribution has been that another car has been preserved for posterity and that he has shared his interest with the thousands of others so dedicated.

A car, whether it be in the passing parade or on the fields of competition, stands on its own merits regardless of who is the owner, or who has restored it. For example, the Horseless Carriage Club in its directions to judges instructs: "Owner restoration should be encouraged but not given preference. It is impossible for an owner to completely restore every part of his car."

Do It Yourself Can Save Money

It is understandable that to perfectly restore a car many others must lend their assistance. Such help often comes voluntarily and without cost from fellow club members. But still, much work must be done by the day-to-day professional craftsman. But a great deal of money can be saved by the hobbyist who can do as much work as his time and inclination permits.

For example, the tremendous cost of labor to completely remove every nut, bolt, part, clean it, sandblast it, buff it, and put a primary coat on it is often 60 to 70 percent of the total cost of restoration. Work thus done by the individual easily balances out the cost of such necessaries as, for example, four new wheels completely built, a new casting, or a gear machined.

Whether you do it yourself to save money, to relax, or to prove the challenge that you can do it when you total up the work done, there will be a great number who have helped make your dream come true and made possible a car that is the pride and joy to all.

Collecting car emblems has become a hobby in itself. Many collectors of emblems can provide the missing emblem to what would have been an incomplete restoration. The cars of today seem to have lost the colorful touch in emblem design compared with cars of yesterday

WHEN WAS IT MADE?

Like the present day Volkswagen, the Model A underwent a steady refining process in mechanical and trim details while remaining substantially the same in outward appearance. From month to month in the lifetime of the car certain changes were made, some minor, some of major importance to performance and service. In restoring an A, while it is desirable to replace each mechanical component with its original counterpart, only a fanatic would insist, for example, that the unsatisfactory multiple disc clutch be used, (if located) when the later single plate type could be installed. These parts are interchangeable, in fact most owners of the early 1928 models went back to their dealer and arranged for a swap when the newer clutch become standard. However, there are certain authentic touches that the restorer will strive to maintain, such as the electric windshield wipers used before August 1929.

To help you determine what components were standard on your A, the following highly interesting chronology of the development of the car is reprinted from HENRY'S FABULOUS MODEL A FORD, By Leslie R. Henry published by Floyd Clymer.

"Revolution" was Henry Ford's awesome term for his disturbing experience of change-over from Model T to MODEL A. He never expected to repeat it; rather, he expected MODEL A to remain the standard Ford product fuly as long as did Model T. But what Henry Ford did not realize was that, while Model T, at creation, had been nearly twenty years ahead of its time, MODEL A was actually but a timely compromise, albeit a good one. That it continued as a successful—even a formidable—contender in the changing automotive world for those four years was a tribute more to its dependability and stamina than to its modernity. There was yet another factor favoring MODEL A then, its high quality at low price. This was the answer for the man caught in the economy squeeze of the "great depression" but who still wanted a good, *new* car.

Henry Ford knew that in the continued production of MODEL A he would have to make many small and gradual alterations for, after all, the car had gone directly from the drawing room into production. These expected changes he called "evolution." Evolution never disturbed him nor his plant; evolution he could control and turn to advantage; evolution enabled him to effect improvements and economies. And evolve is just what MODEL A did.

What interests us now are some of the more relevant of the nearly 5,000 minor and the 150 major changes made to MODEL A from its beginning on October 21, 1927 to its end on April 30, 1932—changes which in number fell just short of equaling MODEL A's 5,500 parts!

So standardized was MODEL A that only one major body style change was made, and that in the 1930 Fords. Ford officially recognized no change between the 1928 and 1929 cars—but the changes were many and minute. Ford did recognize a change between the 1930 and 1931 cars—but the changes were few and faint. Consequently, a passing MODEL A can only be identified *quickly* as either the 1928-29 style (SEE FIG. 46)—as characterized by its low hood and nickel radiator shell, its reverse curved cowl and coupé pillar, and its 21" wheels—or the 1930-31 style (SEE FIG. 44)—distinguished by its high hood, its stainless steel radiator shell, it smooth, tapered cowl, and its 19" wheels with the larger hub caps. More positive identification requires closer examination; most of the prominent, *general* features peculiar to each style year are summarized briefly here, then considered in more detail:

Photo courtesy Ford Motor Company.

FIGURE 44. THE FAMILY SEDAN IN 1930.

This Tudor Sedan, 55-B, was the most popular of the entire Ford line, particularly as a "family sedan." With the windshield open, as shown, air not only came directly in but was directed down to the floorboards by the cleverly shaped interior cowl panel or "facia board."

MODEL A FORDS — SOME PROMINENT GENERAL CHARACTERISTICS OF EACH STYLE YEAR

	1928	1929	1930	1931
Head Lamps	Nickel, "acorn".	Nickel, "acorn", "Twolite".	Stainless, parabolic, "Twolite".	Same.
Head Lenses	Vertical flutes.	Vertical flutes with prisms.	Same.	Same.
Tail Lamp(s)	Nickel, cylindrical, "Duolight", forged bracket.	Nickel, cupped, "Duolamp", pressed steel bracket.	Stainless, cupped, "Duolamp", pressed steel bracket.	Same.
Radiator Shell	Nickel, low & rounded, 13/16" wire holes, teardrop tab on crank hole cover, blue enamel emblem.	Nickel, low & rounded, 15/16" wire holes, dumbbell tab on crank hole cover, blue enamel emblem.	Stainless, high, painted panel insert at bottom only. Blue enamel emblem.	Stainless, high, painted panel insert at bottom and at top. Pressed stainless emblem.
Bumpers	Nickel, reverse curved ends, **round** head on center bolt*.	Nickel, reverse curved ends, **oval** head on center bolt*.	Chromium, slight bow, all bolts capped with pressed stainless*.	Same.
Cowl	Reverse curved with coupe pillar, exposed fuel tank.	Same. (Except for Briggs & Murray bodies.)	High, tapered smoothly into hood, concealed fuel tank.	Same.
Steering Wheel	Red plastic, dished.	Black plastic, dished.	Black plastic, flat, large hub.	Same.

MODEL A FORDS — SOME PROMINENT GENERAL CHARACTERISTICS OF EACH STYLE YEAR

	1928	1929	1930	1931
Dash Panel	Nickel, heart shaped. Oval speedometer.	Same.	Same. (Mid-year change to 1931 style.)	Stainless, oval, ribbed center, round speedometer.
Wheels & Tires	21 x 4.50.	Same.	19 x 4.75.	Same.
Running Boards	Separate, ribbed rubber, zinc trim.	Same.	**Integral** with splash apron, pyramid rubber, stainless trim.	**Separate,** pyramid rubber, stainless trim.
Splash Aprons	One piece.	Same.	Two piece.	One piece.

*Note that all bumper bolt heads had the depressed portions painted the same shade of blue as the porcelain enamel radiator emblem regardless of body color.

Photo by Utzy, Jenkintown, Pa.

FIGURE 46. "A GOOD JUDGE OF OLD CARS."

Judge George C. Corson of Plymouth Meeting, Pa., a vice president of AACA and long a MODEL A enthusiast, recently restored this beautiful 1929 Ford Tudor Sedan, 55-A. Judge Corson took the car completely apart for cleaning, inspection, painting and nickel plating but, since it had been driven only 14,000 miles, no mechanical replacements were required to restore it to first class condition.

January, 1928.

About 5,000 MODEL A Fords had been built with the single brake system before threats of adverse state legislation forced Henry Ford to introduce the separate, extra parking brake equipment. This major change resulted in completely obsoleting the original "AR" type brake drum, backing plate, hub, wheel, wheel lugs, and hub cap. The new rear brake drum was changed from a single to a dual drum and the new "B" type wheel was made with a larger lug plate to fit over the new "B" type rear brake drum. The old style hub cap, which closely resembled the Model T cap with the hexagonal bead around the outer edge, was enlarged and the outer bead made circular. The Model T type forged wheel lug, which had a tapered shoulder larger in circumference than the hexagonal shank itself, was changed to the less expensive lug we know today. This new lug was simply turned from "hex" stock to make a rounded head and a tapered bearing surface against the lug plate of the wheel.

The new "B" type wheels and drums then remained unchanged for the duration of the MODEL A; it is important never to interchange the "AR" and "B" wheels and drums because of the resulting improper supporting of wheels on mismated drums.

With the change in the brake system, the hand lever was relocated from the left of the frame to a position directly in front of the gear shift lever. The opening thus left in the side of the body floor was covered by a plate at first screwed into place, then riveted in place. By mid-year, the hole was eliminated.

Earliest of the MODEL A Fords had the old Model T type bumpers without an end bolt; these were soon changed to the familiar type.

The Phaeton and Roadster (and Roadster Pick-up) were first fitted with outside door handles about this time.

February, 1928.

While the heavy trucks were still designated as MODEL AA, the stamping of the "AA" prefix on the engine block numbers was discontinued with the adoption then of the same type clutch spring in all engines. Thereafter, all engine numbers had only the "A" prefix.

Photo by Duane Fielding, Prospect Park, Pa.

FIGURE 40. A LATE 1928 FORD PHAETON.

Built late in 1928 (Serial No. A650684), this Phaeton was factory equipped with the improved single disc clutch. All other 1928 features are unchanged: it has the cylindrical tail lamp, red steering wheel, light brown upholstering, "power house" generator, hand brake (squeeze type) directly in front of gear lever, the little round Ford escutcheon bolt in center of front bumper, and the vertically fluted head lamp lenses.

Arabian sand body with seal brown mountings, and black fenders, wheels, and running gear are authentic for this car.

The riveting of the splash pan within the oil sump pan was discontinued; the splash pan was then held in place by being snapped into grooves in the sides of the oil pan.

May, 1928.

MODEL A production was at last in full swing at about 8,000 cars per day (the all-time high for MODEL A production was 9,100 on June 26, 1929).

Now the Briggs-built Fordor Sedan body (SEE FIG. 44) was added to the Ford line. It was more luxurious than the other bodies and it eliminated the anular "coupé pillar" by introducing the concealed fuel tank under the tapered cowl—this styling was to become the outstanding mark of all the 1930-31 Fords. The Fordor had nickeled moulding at the juncture of hood and cowl, and the cowl itself had a ventilator door in the side. The rear upper quarter was "blind" and covered with leather while the top was a Seal Brown fabric.

The Business Coupe was a new body which resembled the Sport Coupe with a fabric covered rear quarter but without the Landau irons of the Sport Coupe.

While early Ford catalogs showed a four door Ford having the coupé pillar and reverse curved cowl just like the Tudor model, this body style was never produced as a Sedan, but it did appear later as a taxicab.

Genuine accessories introduced this month were windshield wings and top boots for the open cars.

July, 1928.

During this month two new makes of "ah-ooga" horns were added to the original Sparton—the E. A. and the Ames. For a while these had three different horn motor covers; later all were standardized.

More genuine Ford accessories included a spare wheel lock, a Boyce motometer bearing the Ford emblem, and the first of the "quail" radiator cap ornaments. The quail cap later became the mark of the 1930-31 Ford Roadsters and Phaetons in particular.

> Luther H. Killam of Durham, Conn., uncovered the story behind the "winged quail" ornament. In a letter to Luther, Glen A. Johnson of the Stant Manufacturing Company, Incorporated, of Connersville, Ind., stated:
>
> "It might interest you to know that the writer designed and modeled this ornament as one of his first projects for Stant. We secured live quail from the Indiana Conservation Department, and built special padded cages for them as they beat themselves to death against wood or metal screening. We photographed them in all positions and measured them to get complete records before returning them to the Conservation Department. We made models in clay, wax, and finally brass from which Ford gave approval. Diecasting dies were then engraved to match exactly the master brass model and hundreds of thousands were run from these dies.
>
> The ornament was produced as a service item for a number of years after production of MODEL A was stopped. An added note and the part we do not like is the fact that Ford ordered these dies scrapped a number of years ago."

August, 1928.

Landau irons were made available as accessories for imparting a "sporty look" to the Business Coupes already on the road, and detailed instructions for installing the irons were furnished all dealers.

The Briggs Body Company was by this time supplying about 2,500 bodies daily to Ford and the Murray Body Company was just beginning to fill Ford orders for more bodies.

A "Special Coupe" was introduced; this was exactly like the first coupe (45-A) except the Special Coupe (49-A) had the rear top quarter covered with black leather.

September, 1928.

The original two-piece venturi tube in the Zenith carburetor was changed to a one-piece tube.

A lubrication fitting was added on the top of the steering sector shaft near the frame of the car.

The original type pointer with numerals were omitted from the shock absorbers.

October, 1928.

Henry Ford's troublesome Abell starter drive with its flimsy ½" drive shaft was obsoleted and succeeded by the standard Bendix drive having a ⅝" drive shaft. These two starter mechanisms were not interchangeable in any way.

The removable, round plate in the bottom of the oil pan under the oil pump was eliminated.

November, 1928.

The five-bearing cam shaft was replaced by a three-bearing shaft and the engine block was altered accordingly. These two different cam shafts were interchangeable in either block.

The red plastic steering wheel was replaced by black plastic which didn't rub off onto the driver when wet, as did the red!

The four-point engine mounting was changed to a three point system with a "floating" front supporting yoke. For this the two original upward-projecting tabs were cut off the front cross member and a forged yoke supported on two small springs was substituted. (Later parts suppliers offered very good rubber pads in place of the two springs on top of the cross member and the one spring under it.)

A new solid "oil-less" brake cross shaft was installed to eliminate the original lubricated type—in the interest of manufacturing economy.

At about engine number 560,000, the troublesome multiple-disc clutch (patterned after the Lincoln clutch in all but performance!) and its flywheel were obsoleted by the new, improved single-disc clutch assembly.

December, 1928.

The lower "leg" was added to the choke lever at the butterfly valve to permit attaching a wire to run through the radiator for operating the choke when cranking.

By this time the 5-brush "Powerhouse" generator was no longer furnished on any of the engines. The less troublesome, more conventional 3-brush generator had already become standard for the duration of MODEL A.

The official showing of the 1929 Fords was in January and, though Henry Ford did not announce any "changes" in the 1929 Fords, he did introduce many new body styles.

January, 1929.

Among the new Fords for 1929 was the taxi cab. This has the "coupé pillar" characteristic of the Tudor Sedan and, perhaps, was the design originally intended for the Fordor cars in 1928 but never produced. Of course, the interior was especially dsigned for taxi service; the driver's seat was enclosed by a glass partition and a "jump" seat permitted a fourth person to ride in the rear compartment.

Most striking of the new Fords for 1929 was the Town Sedan (not to be confused with Town Car). This was an exceptionally luxurious four-door, 3-window sedan; the interior was quite roomy, the rear floor was dropped in the center, the upholstering was in rich Mohair with side arm rests and a pull-down center arm rest in the rear cushion. Nickeled cowl lamps were "standard" equipment, with trunk, trunk rack, and bumper extensions as optional equipment. Both Briggs and Murray built these bodies; the Murray-built Town Sedans had white metal cowl lamp brackets, while the Briggs had brass castings.

The Town Car was offered for those who wanted a car with both "snob appeal" and maneuverability and economy in afternoon down-town transportation. The chauffeur usually sat in the open although a snap-on fabric cover was provided for him.

The Station Wagon (built by Murray) was introduced later in the year and was another Ford "first" in that it was the first station wagon in the industry to be assembled in the factory; all other station wagons were more or less "custom built" on commercial chassis by independent body builders.

THE 1929 MODEL A FORDS

Body Model	Body Name	Price
35-A	Standard Phaeton	440
40-A	Standard Roadster	435
45-A	Standard Coupe	500
49-A	Special Coupe	510
50-A	Sport Coupe	530
54-A	Business Coupe	490
55-A	Tudor Sedan	500
60-A	Fordor Sedan (Briggs) Leather Back, Seal Brown Top	600
60-B	Fordor Sedan (Briggs) Leather Back, Black Top	600
60-C	Fordor Sedan (Briggs) Steel Back	600
68-A	Cabriolet	645
135-A	Taxi Cab	800
140-A	Town Car	1200
150-A	Station Wagon	
155-A	Town Sedan (Murray)	670
155-B	Town Sedan (Briggs)	670
165-A	Standard Fordor Sedan (Murray) (3-window)	625
165-B	Standard Fordor Sedan (Briggs) (3-window)	625
170-A	Standard Fordor Sedan (2-window)	625

(Commercial line not listed here.)

In spite of the many major engine changes made two months earlier, a few more were effected. A new, solid-skirt, light weight aluminum piston was adopted to replace the original split-skirt aluminum piston.

The timing gear cover plate was simplified by elimination of an external rib and the "timing pin" was changed from a hexagonal section to a square section.

The long, oval handle on the oil dip stick was changed to a small circular handle.

The starter switch pedal was made smaller and was screwed into the switch.

Finally, the shroud behind the radiator and around the fan, which added so much to the efficiency of the fan, was eliminated as a matter of economy. Many people mistakenly believed that the shroud reduced the flow of air through the radiator!

Advent of the 1929 models marked the beginning of a progressive replacement of the forged body and fender and lamp brackets with pressed steel brackets. Until June of 1929, there was often a mixture of both forged and pressed steel running board brackets on the same car, for instance!

February, 1929.

New "Twolite" head lamps were introduced this month along with the cup-shaped "Duolamp." The original one-bulb (21-3 CP) head lamps (parts A-13004-A and A-13005-A) were succeeded by either a two-bulb head lamp (A-13005-C) for body styles without cowl lamps, or a single-bulb type (A-13005-D) for cars with cowl lamps. The Twolite lamps had new lenses with prisms added to the former fluted type.

A new lighting switch was provided having an extra contact for the parking light bulb in the Twolite head lamp or for the cowl lamps on the Town Sedan.

The holes in the radiator shell were enlarged from $13/16$" to $15/16$" to permit the new three-wire cables and enlarged terminal plugs to pass through.

A percentage of the 1929 Fords were now being assembled with a new two-tooth steering gear sector which was fully adjustable for wear. Many months passed before the original seven-tooth gear sector assembly was completely obsoleted.

March, 1929.

The front wheel hubs were changed from expensive forgings to the pressed steel type; these remained standard for the duration of MODEL A.

Cowl lamps were offered through the dealers as accessories for installation on Fords not regularly equipped with them.

The flywheel web was increased in thickness from ⅜" to $^{25}/_{32}$" where it mated with the crankshaft flange; longer cap screws were required thereafter for this flange.

April, 1929.

The rear main crankshaft bearing cap was changed from a forging to a casting, which had to be made larger to obtain necessary strength. The cap bolts were then lengthened accordingly from 3⅜" to 4³⁄₁₆". This change was one for manufacturing economy, as were all such changes from forgings to castings or stampings.

The fuel valve handle under the cowl tank was also changed from a forging to a pressed steel handle.

The entire throttle assembly (A-9725) was redesigned for simplicity and economy; the accelerator pedal shaft was changed from two- to one-piece construction (A-9734).

May, 1929.

To reduce tendency for engine to burn oil; the valve chamber cover was changed to lower the oil return pipe and thus lower the oil level in the valve chamber.

Spark plug gap specification was changed from .027" to .035".

June, 1929.

The new three-window Fordor Sedan, model 165-A and 165-B, was introduced this month. This model resembled the Town Sedan but had no cowl lamps nor luxurious interior.

The universal joint housing cap was changed by removing the extension lug which had served to support the torque tube on the center cross member during assembly before mounting the engine in the chassis.

July, 1929.

The engine valve guides were shortened from 2⅜" to 2⅛" to help prevent valve sticking.

The clutch release-arm fastening pins were increased in diameter so they would be strong enough to permit elimination of the former keys and keyways—this reduced manufacturing costs.

The cylinder block rear wall was reinforced inside the block.

A new breather pipe was installed on the crankcase which had four baffles sloping upward instead of downward in an attempt to reduce oil blow-by.

An oil hole was made in the starting crank bearing to permit lubricating the front spring within the cross member.

A second type of head lamp plug and socket (A-14584-C and A13075-B) were introduced which obsoleted the original part (A-13075-AR).

The hand brake lever was moved from its position in front of the gear shift lever to the right side of the gear shift lever.

August, 1929.

Electric windshield wiper motors were replaced by the vacuum type starting in the Tudor Sedans and gradually extending to all body types by July, 1930. The intake manifold was now drilled and tapped for a vacuum hose connection.

October, 1929.

A larger water passage was provided in the center of the cylinder block, head, and head gasket. The new type head gasket could be used on the older engines.

November, 1929.

The oil pump body was changed from a forging to a casting and the shank was then ribbed for strength. The

THE 1930 MODEL A FORDS

Body Model	Body Name	Price
35-B	Standard Phaeton	440
40-B (Std.)	Standard Roadster	450
40-B (De Luxe)	De Luxe Roadster	520
45-B (Std.)	Standard Coupe	495
45-B (De Luxe)	De Luxe Coupe	545
50-B	Sport Coupe	525
55-B	Tudor Sedan	495
68-B	Cabriolet	625
140-B	Town Car (Same as in 1929)	1200
150-B	Station Wagon	640
155-C	Town Sedan (Murray)	640
155-D	Town Sedan (Briggs)	640
165-C	Standard Fordor Sedan (Murray)	600
165-D	Standard Fordor Sedan (Briggs)	600
170-B (Std.)	Standard Fordor Sedan (Briggs)	600
170-B (De Luxe)	De Luxe Fordor Sedan (Briggs)	640
180-A	De Luxe Phaeton	625
190-A	Victoria Coupe	580

pump shaft and brushing size was increased from $2\frac{1}{32}"$ to $\frac{5}{8}"$.

The front fender pressed steel bracket was changed to give more tire clearance. Only two vertically-spaced bolts and holes were used for attaching the new bracket instead of the former three.

January, 1930.

January again was the month for showing the new Ford models, and this time Henry Ford proudly announced there were major changes in the MODEL A. These were mostly in the new bodies which were better proportioned with higher hoods and radiators and with smoother cowls and body panels. Other changes included the use of chromium and stainless steel for radiator shell, lamps, and trim; new, larger "balloon" tires, size 4.75 x 19; shallower, "parabolic" shaped head lamp shells; slightly bowed bumper bars (chromium plated); and the flat steering wheel with a large hub. By this time, all the steering sector gears were of the two-tooth type and were fully adjustable for wear. Spark and throttle control rods were contained entirely within the steering column.

The radiator shells were stamped from stainless steel; because of the difficulty in deep-drawing the stainless steel and also for economy, the bottom panel was crimped in place and was painted instead of being highly polished.* The original style blue enameled radiator emblem remained, but the old style radiator cap with fluted edges and internal threads was replaced by a flat cap with knurled edge and with a quarter-turn interrupted screw for fastening. The fuel tank cap was also changed to match. Fenders were wider, lower and more graceful.

February, 1930.

The metal tube running from the terminal block to the generator cut-out as a conduit for the wires was changed to a pliable black lacquered loom. This was done

---

*Joe Galamb, the designer, stated that the inserts were made of carbon steel to save the cost of stainless steel. Of course, carbon steel *had* to be painted to keep it from rusting.

because of complaints about breakage of the metal tube and because of frequent chafing of the wires and subsequent short-circuiting on the conduit.

The steering column was increased 1" in length and the column bracket was shortened ⅜" to provide easier steering and handling.

March, 1930.

The rear main bearing cap oil pipe was enlarged from $5/16$" to ⅜" to permit quicker drainage of the oil in cold weather.

Rear engine supports were redesigned and made of heavier gauge pressed steel. Bolts used with the new supports were $1^{17}/_{32}$" long instead of 1⅛" as formerly.

The safety screen in the fuel tank hole was also made of heavier gauge metal and supplied with slots ½" deep (instead of only ⅜") for removing the screen without danger of it breaking.

The "stack" strap, used to hold the top in place when folded down, was now supplied as regular equipment on all roadsters.

Brake drums were improved by being rolled true and to finished dimensions instead of being ground to dimension as formerly. Rolling the drums produced a "worked hardened" surface which resists wear better than a ground surface. After this time, Ford recommended that customers buy new genuine Ford drums rather than having the old, worn drums remachined.

The former means for adjusting the parking brakes was discontinued; this saved manufacturing cost and Henry Ford believed parking brakes never needed adjustment in service.

Steering gear ratio was increased from 11¼ to 1 in the old gear to 13 to 1 in the new gear for easier turning of the steering wheel.

Artificial leather (cross Cobra grain) became optional upholstering material (trimming) for the Coupe and was suggested as being particularly attractive for Coupes used "for business purposes."

Brown mohair or deep tan Bedford cloth became the new optional upholstering material in the De Luxe Coupe, De Luxe Sedan, and the Town Sedan.

A dome light was now furnished in the De Luxe Coupe.

Spark plug gap specifications were increased from the range of .025" to .030" to the new range of .027" to .035".

April, 1930.

The generator shaft and pulley was changed by shortening the shaft and the hub. Old pulley hub length was 1⁵⁄₃₂"; new hub became ⅞" long.

The emergency (parking) brake cross-shaft was changed from tubular to solid section.

The horn mounting bracket was widened from 1" to 1⅛" to fit the new head-lamp tie rods. The new bracket can be used on the old style rods with the "acorn" shaped head-lamps, but not vice versa.

May, 1930.

The breather pipe on the crankcase was changed for a third time to a pipe having three internal disc baffles with their centers bent downward, alternately left and right. This was the last attempt to check oil blow-by. (Owners solved this problem by installing an accessory flexible breather pipe to conduct oil vapors down below the car frame.)

Lubrication fittings were added to the water pump bushings and to the brake and clutch pedals.

The front shock absorber arms were offset to provide more clearance between arm and body. These new arms are interchangeable with the older style.

June, 1930.

A new body type was added: the De Luxe Phaeton. Having but two doors, a lowered steering wheel, and a lower, chromium plated windshield frame, this car presented a much sportier appearance than the four-door Standard Phaeton. The front seats folded forward to give access to the rear seat; all upholstering was in genuine leather with a two-tone grained effect. Regular equipment included cowl lamps; folding, chromium-trimmed rear trunk rack; one spare wheel mounted on the left side; rear view mirror on the windshield post;

and a chromium-plated windshield wiper motor, *now vacuum operated on all open body types.*

Because of the lowered steering column, a bend was put in the accelerator-to-steering column rod (A-9742-B) to allow it to clear the starter push rod.

The familiar heart-shaped, satin-finish instrument panel with the oval speedometer was changed to a horizontal oval stainless steel panel having a ribbed center and a round speedometer.

A two-piece metal spare tire cover was offered as an accessory; this was enameled black and was available either with or without chromium trim.

August, 1930.

The De Luxe Roadster appeared this month as the fourth open body type in the Ford Line. Regular equipment on this was the same as on the De Luxe Phaeton except the rumble seat cushions which were upholstered in artificial leather to match the appearance of genuine leather of the front seat, and there was a spare wheel mounted on each side of the car. This was the only MODEL A Ford to have twin side-mounted spare wheels as regular factory equipment; all other twin mountings were either on special order to the factory or were added by the dealers after delivery.

A new emergency brake lever assembly with finer ratchet teeth became standard for all Fords; this permitted finer adjustment of parking brake tension and easier release of the brake handle.

Bumper bars were reduced in length from $62\frac{7}{8}"$ (A-17757-C) to 60" (A-17757-D). This dimension is measured as the chord of the bumper arc, not along the face of the bumper bar.

Valve spring (A-6516-A1) was changed from $3\frac{7}{16}"$ free length to $2\frac{15}{16}"$ free length (A-6513-A2).

November, 1930.

The graceful Victoria Coupe, with the "bustle back" (which was luggage space reached from behind the rear seat cushion) was another completely new body type.

THE 1931 MODEL A FORDS

Body Model	Body Name	Price
35-B	Standard Phaeton	435
40-B (Std.)	Standard Roadster	430
40-B (De Luxe)	De Luxe Roadster	475
45-B (Std.)	Standard Coupe	490
45-B	De Luxe Coupe	525
50-B	Sport Coupe	500
55-B	Tudor Sedan	490
68-B	Cabriolet	630
68-C	Cabriolet	630
150-B	Station Wagon	625
155-C	Town Sedan (Murray)	630
155-D	Town Sedan (Briggs)	630
160-A	Standard Fordor Sedan	590
160-B	Town Sedan	630
160-C	De Luxe Fordor Sedan	630
165-C	Standard Fordor Sedan (Murray)	590
165-D	Standard Fordor Sedan (Briggs)	590
170-B	De Luxe Fordor Sedan (Briggs)	630
180-A	De Luxe Phaeton	580
190-A	Victoria Coupe	580
400-A	Convertible Sedan	640

Added this month to the Ford line, it was the first closed MODEL A to have a slanting windshield, to have no sun visor, and to have a *deep* drop in the floor to permit a lowered roof and silhouette. The Victoria Coupe bodies having a fabric-covered rear top quarter were made by the Briggs Company; those with the steel rear top quarter were made by Murray.

These three new bodies, the Victoria Coupe, the De Luxe Phaeton, and the De Luxe Roadster, were brought out by Ford in an attempt to capture those customers who normally would have purchased medium- or high-priced cars but whose income or budget had shrunk with the deepening of the current economic "depression."

Reversing the former trend, Ford changed the front fender brace from a steel stamping to a forging because of fatigue cracking of the pressed steel bracket where it was bolted to the car frame.

For the 1931 MODEL A, Henry Ford claimed no change in models. As we have already noted, the only apparent changes involved merely the addition of a painted panel inserted at the top of the stainless steel radiator shell, substitution of a pressed stainless steel emblem for the former enameled plaque; and a return to the separate running board and the one-piece splash apron. Many of the 1931 sedans appeared this year with a visorless windshield.

On December 7, 1931, Henry Ford's revolutionary "en bloc" V-8 engine in Edsel Ford's new body styles appeared as the new Ford line for 1932. But MODEL A was continued in production until April 30, 1932, when the Model B Ford was announced.

Model B had the graceful body of the all-new V-8 Fords, but retained the Model A engine modified by the addition of a camshaft-operated fuel pump and by the addition of a centrifugally operated spark advance in the distributor. Following the Model B engine came a Model C which found application in several commercial chassis in the 1933 Ford line.

Model A Motor Numbers

Month	First No.	Last No.
1927		
October 20, 1927	1	137
November	138	971
December	972	5275
1928		
January	5276	17251
February	17252	36016
March	36017	67700
April	67701	109740
May	109741	165726
June	165727	224276
July	224277	295707
August	295708	384867
September	384868	473012
October	473013	585696
November	585697	697829
December	697830	810122
1929		
January	810123	983136
February	983137	1127171
March	1127172	1298827
April	1298828	1478647
May	1478648	1663401
June	1663402	1854831
July	1854832	2045422
August	2045423	2243920
September	2243921	2396932
October	2396933	2571781
November	2571782	2678140
December	2678141	2742695

Model A Motor Numbers (Continued)

Month	First No.	Last No.

1930

January	2742696	2826649
February	2826650	2940776
March	2940777	3114465
April	3114466	3304703
May	3304704	3509306
June	3509307	3702547
July	3702548	3771362
August	3771363	3883888
September	3883889	4005973
October	4005974	4093995
November	4093996	4177733
December	4177734	4237500

1931

January	4237501	4310300
February	4310301	4393627
March	4393628	4520831
April	4520832	4611921
May	4611922	4695999
June	4696000	4746730
July	4746731	4777282
August	—	—
September	4777283	4824809
October	4824810	4826746
November	4826747	4830806
December	—	—

1932

January	4830807	4842983
February	4842984	4846691
March	4846692	4849340

MAKING THE CHANGES

As each innovation was slipped into the production line, a Dealer Service Bulletin covering that change, along with many helpful hints on general service, was issued to Ford dealers. To amplify the listing of changes mentioned in the previous chapter, we are reprinting a selected number of these bulletins.

FORD SERVICE BULLETIN *for January*

SHOCK ABSORBERS

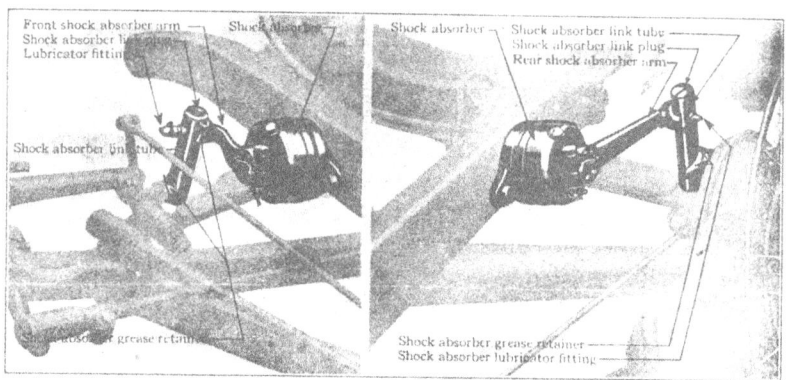

FIG. 417　　　　　　　　　　FIG. 418

Ford hydraulic double acting shock absorbers operate entirely on the principle of hydraulic resistance. Glycerine is forced from one chamber to another by the movement of the lever arm. The working chamber is automatically kept full by the glycerine in the reservoir.

As the shock absorbers are accurately adjusted at the factory, it should not be necessary to alter this adjustment except in rare cases where more or less shock absorber action is desired.

The markings "C. W." (clockwise) and "A. C." (anti-clockwise) are stamped on the side of the reservoir of each absorber.

When installing, be sure that the instruments marked "C. W." are installed at the right front and left rear side of frame. Instruments marked "A. C." are installed at left front and right rear.

Adjustment

You will observe a needle valve with an arrow pointer extending through the center of the shaft. Surrounding it numbers from 1 to 8 are stamped. The average setting is with the arrow pointing at 2 for the front shock absorbers and 3 for the rears.

Turning the needle valve changes the adjustment. Resistance is increased when the needle valve is turned from 1 to 8 and decreased from 8 back to 1. A slight movement of the needle valve either way makes a big difference in the action of the instruments. During extremely cold weather it may be found advisable to further decrease the resistance by turning the needle valve back to 1.

Care

Keep ball joints well lubricated.

Important. The shock absorber arm clamp bolt nut must be kept securely tightened at all times. Make this a part of your regular inspection when cars are in the shop.

The filler plug in the reservoir should be removed at intervals of 5,000 to 10,000 miles, and the reservoir filled with glycerine (Commercial). NEVER REPLENISH WITH OIL. Oil will solidify in the winter or reduce resistance and will not mix with the glycerine in the instruments.

In warm climates, replenish with glycerine, C. P. or Commercial. All instruments contain glycerine with 10% alcohol. Where temperatures of zero and below are prevalent, add an additional ½ to 1 ounce of alcohol.

Lubricating Shock Absorber Connections

The ball joint is made in unit with the instrument arm. It is hardened and ground. The ball joint seats are enclosed in the shock absorber connecting links, and should be lubricated every 500 miles with the compressor gun.

In order to secure maximum riding comfort, it is important that the spring hangers be free in the bushings and kept well greased.

FORD SERVICE BULLETIN *for January*

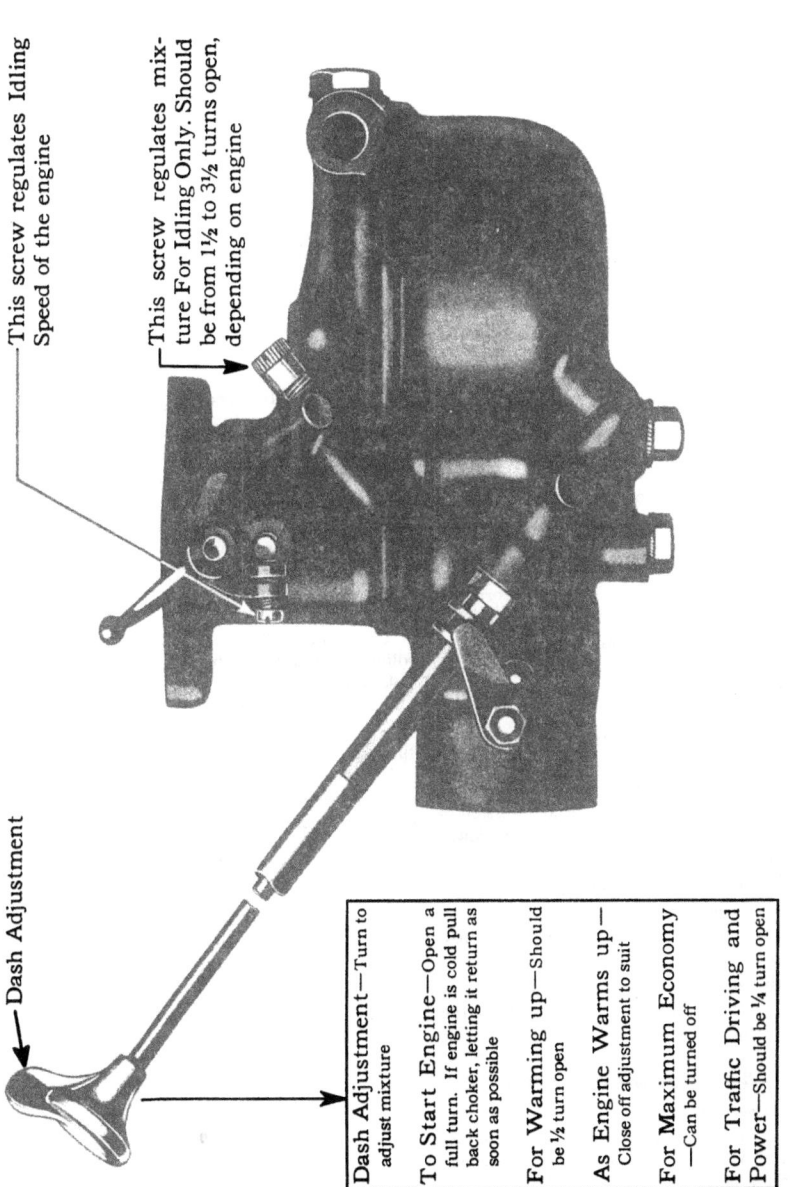

Fig. 419

Model "A" Carburetor Adjustment

Ford Service Bulletin *for January*

Servicing Model "A" Carburetor

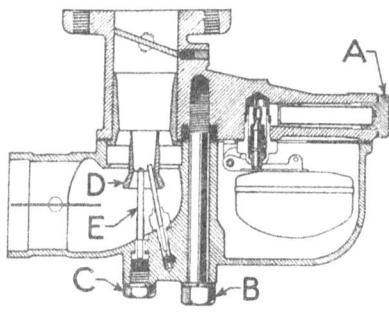

Fig. 420

In cases of suspected carburetor trouble or complaints of poor fuel economy, first check spark plugs, breaker points, compression, etc., before removing carburetor. Many so called carburetor troubles can be traced to one or more of the following causes:

Dirty spark plugs; points incorrectly spaced—Clean points and set gaps to .025".

Breaker contact points burnt or pitted—Dress points down with an oil stone and set gap at .015" to .018".

Leaky manifold · or carburetor connections—With engine idling slowly, flow a little oil on each joint. If engine picks up speed there is a leak.

Poor compression—Check compression in each cylinder by turning engine over slowly with hand crank.

Brakes dragging—Jack up car and see that all wheels revolve freely.

Tires soft—Inflate all tires to 35 lbs. pressure.

If the above points are OK and there is a free flow of fuel through the line, check the carburetor.

Cleaning the Carburetor

Remove filter screen. Blow out any dirt with air or rinse screen thoroughly in gasoline. The screen is easily removed by backing out the filter plug. See "A," Fig. 420. Usually cleaning the screen is sufficient to overcome the trouble.

For complete cleaning, remove carburetor and disassemble it by removing main assembly bolt "B." See Fig. 420. Separate the parts carefully to avoid damaging the gasket, float and idling jet tube.

Remove brass plug "C" beneath main jet, and rinse carburetor bowl in gasoline or use air to blow out any dirt which may have lodged in the bottom of the bowl or in the jets.

Trouble Shooting Hints

Make certain there is gasoline in the tank and a free flow of fuel through the line.

See that the secondary venturi is right side up as shown at "D," Fig. 420.

On complaint of lack of speed, see that main jet "E" is free from dirt.

A plugged compensator, "F," Fig. 421 will result in poor idling and low speed performance.

The idling jet "G" furnishes all the fuel for idling, consequently the tube and metering hole must be kept clear.

In case of leaks see that all connections and jets are tight. If damaged, replace float or fuel valve assembly.

On complaint of poor fuel economy make certain owner understands proper operation of dash adjustment.

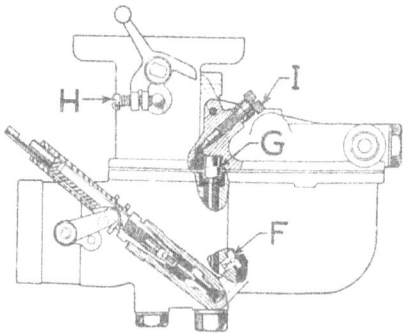

Fig. 421

Water in the fuel line may freeze in cold weather and stop the flow of fuel—use hot cloths for thawing.

The carburetor is a delicate instrument and should be handled carefully. Don't use strong-arm methods in taking it apart, reassembling or handling the various parts. With reasonable care the carburetor will last indefinitely.

Adjustments

Do not expect a new engine that is too stiff to "rock" on compression when stopped, to idle well at low speed.

To Adjust the Idle—If engine is free, fully retard spark lever. Adjust throttle plate adjusting screw. See "H," Fig. 421, so that engine will run sufficiently fast to keep from stalling. Turn idle adjusting screw "I" in or out until engine runs evenly without "rolling or skipping," then back off throttle plate adjusting screw until desired engine speed is obtained. (Make adjustments with engine warm.)

Usually best idling will be obtained with the adjusting screw approximately two turns off its seat.

Dash Adjustment—The dash adjustment does not control the entire fuel supply. A minimum of fuel is constantly drawn from the float chamber through small fixed openings even when the dash adjustment is fully closed.

For best operation under usual driving conditions, the dash adjustment should be backed one-quarter turn off its seat. Running with the adjustment more than one-quarter turn off its seat may be necessary on new stiff engines, but otherwise this will result in poor economy, carbon and crankcase dilution.

The dash adjustment may be turned less than one-quarter turn off its seat to obtain a lean mixture suitable for high altitudes, high test fuels, or when driving at steady speeds on level roads. Under normal conditions, however, too lean a mixture causes uneven running at low speeds and slow pickup.

Do not force the adjusting needle down on its seat as this will score the parts.

Cold Engine Starting

First: Open hand throttle lever two or three notches. Fully retard spark lever. Turn carburetor dash adjustment one full turn to left.

Second: Turn on ignition. Pull back choke rod at the same time depress starter switch. The instant the engine starts, release choke.

Third: As motor warms up, gradually turn dash adjustment to the right until it is in its normal running position—one-quarter turn off seat when engine is warm.

Starting in Cold Weather

These instructions are to aid starting at low temperatures, especially when battery efficiency is low and the engine does not turn over at starting speed.

First: Open throttle lever two or three notches. Fully retard spark lever. Open dash adjustment one full turn and crank engine two or three times with ignition *off* and choke pulled all the way back. This will fill the cylinders with a rich mixture.

Second: Release choke and turn on ignition. Engine should start on second or third quarter turn of the crank.

Warm Engine Starting

With spark control lever about half way down quadrant and throttle lever advanced two or three notches, turn on ignition and depress starter switch. It is usually unnecessary to use choker when the engine is warm.

IDENTIFICATION OF CARBURETOR PARTS

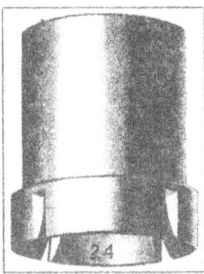

Fig. 422—*Venturi*

The venturi measures the air through the carburetor and keeps it moving fast enough at low speed to completely atomize the fuel.

Ford Service Bulletin *for January*

FIG. 423—*Secondary Venturi*

This is an auxiliary air metering tube which increases the air velocity at the jets to give quick response on acceleration.

FIG. 424—*Main Jet*

This is the long jet. It is connected with the fuel chamber. Its effect is most noticeable at high speeds.

FIG. 425—*Compensator*

The fuel in the bowl flows through this jet into the compensating well. The jet is most effective at low speeds.

FIG. 426—*Cap Jet*

The cap jet controls the rate of discharge from the compensator well into the air stream.

FIG. 427—*Idling Jet*

The function of the idling jet is to measure fuel for very slow running. When the throttle is open, the idling jet is put out of action as the flow of the fuel then changes direction and passes through the cap jet.

GENERATOR CHARGING RATE
Easily Adjusted to Suit Individual Requirements

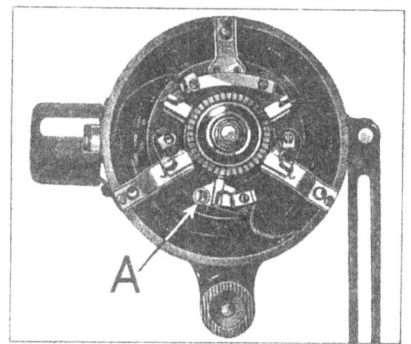

FIG. 428

The generator is mounted on the left hand side of the engine. During winter months the charging rate should be adjusted to 14 amperes; in the summer this rate should be cut down to 10 amperes. The rate can, of course, be increased or decreased to meet individual requirements. For example, the owner who takes long daylight trips should cut down the charging rate to 8 amperes to prevent the battery overcharging. On the other hand, the owner who makes numerous stops, should increase the normal rate if his battery runs down.

Increasing or Decreasing Generator Charging Rate

To increase or decrease the generator charging rate, remove generator cover and loosen field brush holder lock screw. See "A," Fig. 428. The field brush holder can be easily identified, as it is the only one of the five brush holders that operates in a slot in the brush holder ring and which is provided with a locking screw. The remainder of the brush holders are riveted to the ring and are not movable. To increase the charging rate, shift the field brush holder in the direction of rotation; to reduce the rate, shift the brush in the opposite direction. The output of the generator is indicated by the ammeter located on the instrument panel.

ADJUSTING BREAKER CONTACT POINTS

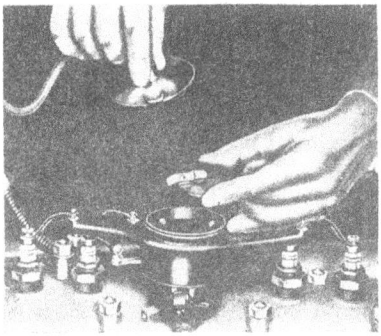

Fig. 429

The gap between the breaker points is set at .015 inch to .018 inch. The gap should occasionally be checked to see that the points are properly adjusted.

If the points are burnt or pitted they should be dressed down with an oil stone. **Do not use a file.**

To adjust the contact points proceed as follows:

Lift off distributor cap, rotor, and body, see Fig. 429.

Turn engine over slowly with starting crank until breaker arm rests on one of the four high points of the cam with the breaker points fully opened.

Loosen lock screw and turn the contact screw until the gap is at .015 to .018. A standard thickness gauge is used to obtain this measurement (see Fig. 430).

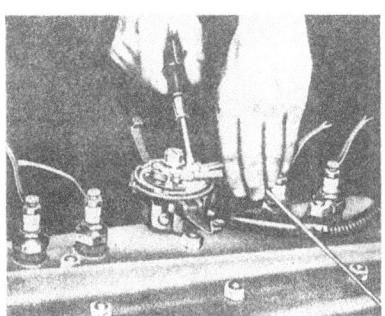

Fig. 430

When correct adjustment is obtained, tighten the lock screw and replace distributor body, rotor, and cap. After tightening the lock screw, it is a good plan to again check the gap to make sure the adjustment was not altered when the lock screw was tightened.

IGNITION TIMING

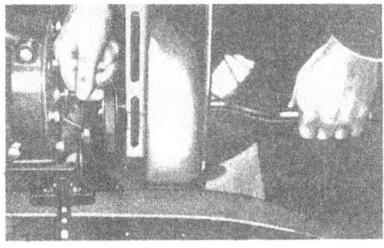

Fig. 431

As the spark must occur at the end of the compression stroke, the timing must be checked from that point. To find the compression stroke and time the spark proceed as follows:

1. Fully retard spark lever.
2. Check gap between breaker contact points and if necessary adjust them as previously described.
3. Screw out timing pin located in timing gear cover and insert opposite end of pin into opening.
4. With the starting crank turn the engine over slowly, at the same time pressing in firmly on the timing pin, see Fig. 431.

Fig. 432

Ford Service Bulletin *for September*

Changes in Model A Carburetor

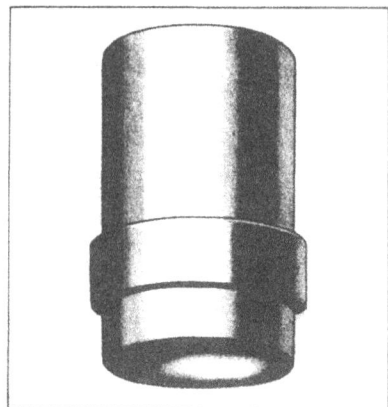

Fig. 564—Single Venturi

Fig. 565—Main Jet

Fig. 566—Compensator

Fig. 567—Cap Jet

Fig. 568—Idling Jet

Several refinements have been made in the Model A carburetor which simplify and add to its smoothness of operation, especially at low speeds.

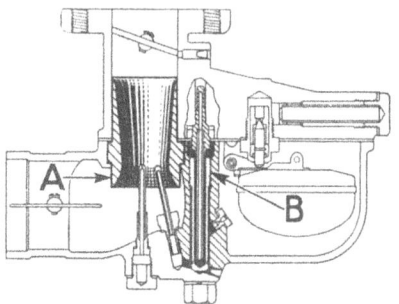

Fig. 569

The original carburetor was provided with a double Venturi made up in two pieces. (See Figures 422 and 423 in the January, 1928, Bulletin.) These parts have been replaced by a longer single Venturi (see A, Fig. 569), the narrowest part of which is 27/32" in diameter. Other changes consist in the addition of a secondary well which is screwed into the lower half of the carburetor as shown at "B" and from which the idling jet derives its supply.

A slight change has also been made in the angle of the throttle plate and the plate is now stamped No. 18½ instead of No. 20. The location of the cap jet in the lower half of the carburetor has also been slightly changed.

These changes necessitated using a different combination of fuel orifices, the parts being stamped as follows:

Main jet now stamped No. 19.5 instead of No. 20.

The compensator is stamped No. 19 instead of No. 18.

The cap jet is stamped No. 21 instead of No. 19.

The idling jet No. 11 instead of No. 10.

The new idling jet is slightly shorter than the old one, the new jet being 3" overall—the old one 3-5/64.

Figures 564 to 568 show the present design parts. Old-style parts are shown in the January, 1928, Bulletin. Never attempt to use old-style parts in present design carburetor. While the parts look alike the fuel orifices in the new parts have been changed to secure maximum results.

The instructions in the January Bulletin regarding carburetor cleaning and trouble checking remain unchanged with the exception that the breaker point gap should be set at .018" to .022" and the spark plug gap .027".

important that all of the plugs be adjusted to a uniform gap of .027".

From letters received it is apparent that some owners still do not understand that after a Model A engine has been run in, the dash adjustment should not remain open more than ¼ turn except for warming up the engine. Owners should be instructed that it not only wastes fuel but it is even harmful to leave this adjustment open longer than necessary.

NEW LUBRICATOR FITTING IN STEERING GEAR HOUSING

The drilled hole in the steering sector shaft, through which the lubricant in the steering gear housing was carried to the sector shaft bushings, has been discontinued. Sector shaft bushings are now lubricated by means of a lubricator fitting placed in the steering gear housing. (See A, Fig. 570.) Lubricant should be forced into this fitting every 2,000 miles.

Should an instance arise where it is necessary to install the present design sector in an old steering gear assembly, it will be necessary to install a lubricator fitting in the housing. This is done by drilling a 21/64" hole 2-3/32" back from center line of flange where housing is bolted to frame. (See Fig. 571.) The hole should be drilled through housing and sector bushing and then tapped with a standard 1/8" pipe tap. When tapping the hole, do not tap too deep. The hole should be tapped just deep enough so that when the lubricator fitting is screwed down tightly approximately two threads on the fitting will remain above the surface of the housing. (See Fig. 571.)

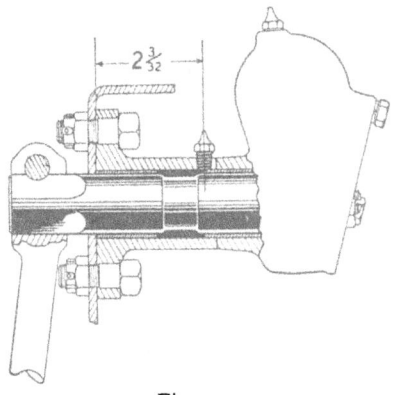

Fig. 571

OIL FOR DUAL HIGH

Use M-533 oil in the dual high assembly. When the assembly is operated at temperatures of 10 above zero or colder M-533 oil should be thinned with 10% kerosene. Approximately one quart of oil is required to bring the oil level up to the "F" (full mark) on the oil level indicator. The oil level should be checked every 1,000 miles and additional oil added if required. Every 5,000 miles the old oil should be drained out and replaced with fresh M-533 oil.

Floor boards used in cars having the original hand brake equipment will not be carried for service.

Should it become necessary to replace floor boards in one of these cars use the present-design floor board together with Floor Board Emergency Lever Plate, part 35245B, to cover the opening in the boards. (See Fig. 572.)

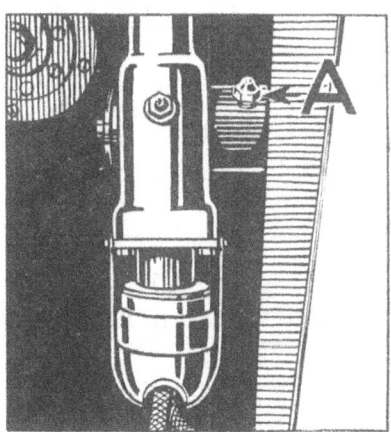

Fig. 570

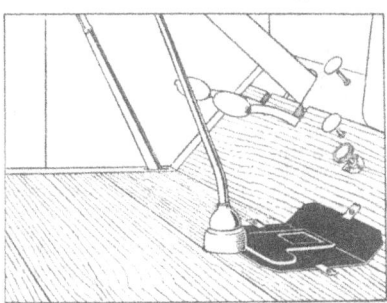

Fig. 572

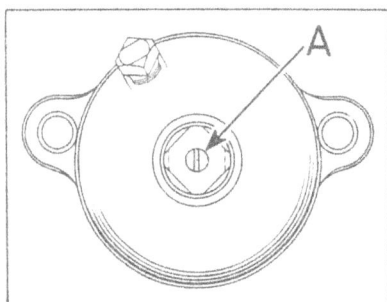

Fig. 573

CHANGE IN METHOD OF ADJUSTING SHOCK ABSORBERS

The pointer on the shock absorber needle valve has been removed and the end of the valve slotted so that it can be easily turned with a screw driver. (See Fig. 573.) The numerals which were stamped on the shock absorber arm have also been removed as they are no longer required.

Slotting the end of the needle valve instead of using the pointer simplifies adjustment and prevents any possibility of the brake rod striking the needle valve.

Adjustment:

Turning the slotted end of the valve changes the adjustment. Resistance is increased when the needle valve is screwed in, and decreased when the valve is backed out.

The average adjustment for rear shock absorbers during warm weather is made as follows: Screw needle valve in until it seats, then back valve off ¼ turn. For front shock absorbers, back valve off ⅜ of a turn.

For cold weather adjustment the needle valve in the rear shock absorbers should be screwed in until it seats, then backed off ½ to ⅝ of a turn. For front shock absorbers the needle valve should be backed off ⅝ to ¾ of a turn.

These settings are of course only approximate and can be easily changed to suit the individual preference of the owner and the conditions under which the car is operated. For example, the owner who drives at high speed over rough roads would require greater shock absorber resistance than the owner who drives at moderate speed over paved highways.

Dealers' mechanics must check owners' cars and adjust shock absorbers to secure maximum riding qualities for the conditions under which the car is operated.

Brake Rod Striking Pointer:

Should you receive a complaint of a brake rod striking the needle valve equipped with the old style pointer, replace the needle valve with one having the new slotted end—this will correct the trouble. The needle valve can be easily removed by screwing it out. In a case of this kind if it is necessary to change one needle valve, all four valves should be changed in order to preserve a uniform appearance and have a uniform method of adjustment.

HUB SHELL CAP A-1130

When installing a hub shell cap, part A-1130, be sure to crimp over 4 of the cap lugs equally spaced around the cap. This will prevent any possibility of the cap coming out of the wheel.

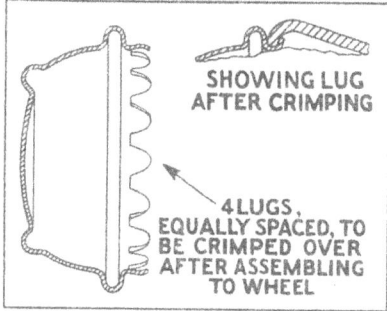

Fig. 574

BEARING A-3123

The outside diameter of the spindle bolt bearing assembly and the steering worm thrust bearing assembly, part A-3123, has been changed to the dimensions shown in Figure 575.

When replacing one of these bearings in the steering gear be sure to install the same design that was removed.

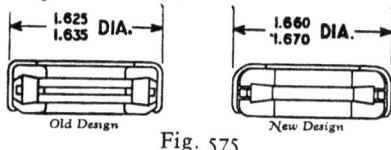

Fig. 575

FRONT WHEEL ALIGNMENT

Front wheel toe-in has been changed from 3/16" to 1/16", plus or minus 1/32". (See A and B Fig. 576.) This new adjustment insures maximum front tire life.

Change in Spindle Arm:

Spindle Arms A-3130-B and A-3131-B have also been changed, i. e., the distance between

FORD SERVICE BULLETIN *for September*

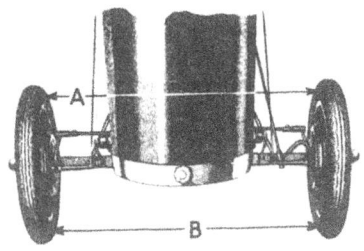

Fig. 576

center line of shoulder (which fits in the spindle) and center line of ball has been changed from 1⅛" to 13/16". (See Fig. 577.) The distance between the shoulder and ball has also been changed from 4 ⅞" to 4 15/16".

These changes in the spindle arm reduce front tire wear to a minimum when turning corners.

To Correct Excessive Tire Wear:

Adjust front wheel toe-in to 1/16" plus or minus 1/32". In practically every case this new adjustment will correct any complaints of premature tire wear. Should an instance arise where it failed to correct the trouble, install the new-type spindle arms.

NEW DECK DOOR SUPPORT

Figure 578 shows the new deck door support now standard in roadster and all coupes without rumble seat.

The new support can be easily installed on cars equipped with the old-design support by proceeding as follows:

Drill a 9/32" hole in deck drain trough

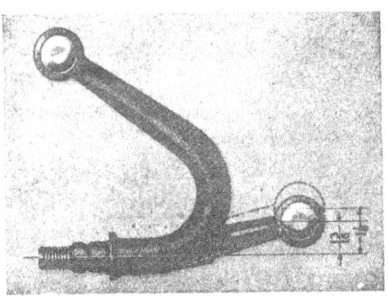

Fig. 577

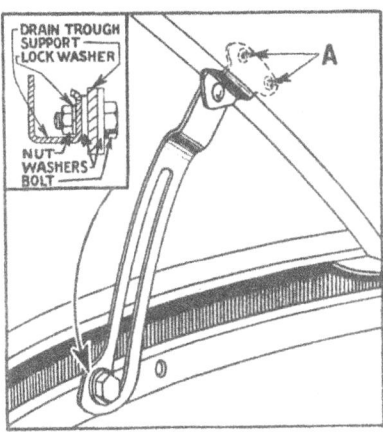

Fig. 578

3-3/32" from center line of rivet "B." (See Fig. 579.) Next remove the two screws "A" and bolt "B" from old style support.

The new support is installed by replacing screws "A" (see Fig. 578) in the same holes from which they were removed and assembling lower part of support to deck drain trough by means of bolt A-20597; one each flat washers A-22266; A-22154; lock washer A-22165 and nut 21668. (See insert Fig. 578.) When bolting new support to drain trough insert bolt through new hole which was drilled.

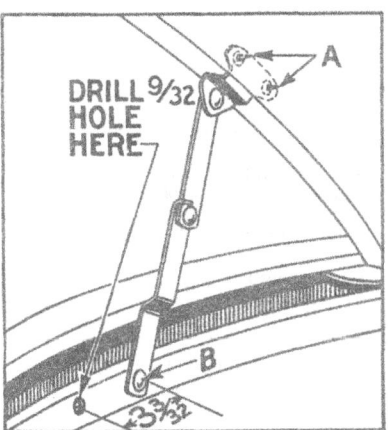

Fig. 579

Ford Service Bulletin *for September*

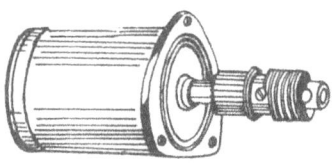

STARTING UNIT A-18475
Fig. 580

NEW SERVICE STARTER DRIVE

We have adopted the Bendix type starter drive which will go into production 100% about October 1st. The Abell type starter drive and parts affected are obsoleted. To take care of service on cars equipped with Abell type starter drives, it is necessary to have a special Bendix type which we are calling "Service Starter Drive."

This new service drive will be assembled to starting motors at branches and dealers can get the complete unit (A-18475) (see Fig. 580) from branches at an exchange price of $3.00 net by returning to the branch, freight prepaid, the old starting motor removed from customer's car. The dealer's price to the customer will be $4.50 net, this price to include installation in car. The starting switch (A-11450-B) which the dealer removes from customer's motor should be placed on the new unit (A-18475) when it is installed in owner's car.

In addition to sending in starter motors removed from customers' cars, dealers will immediately return to the branch their stock of starter motors. These will be reworked at the branch—the new service starter drive installed and the complete unit (A-18475) returned to the dealer at the above exchange price. Dealers must keep a small supply of these built-up units (A-18475) on hand so they can render prompt service to customers. When installing the new unit in customers' cars do not place shims (A-11140-1) between starting motor and flywheel housing.

Before installing a new drive in a customer's car, dealers should carefully inspect the teeth in the flywheel ring gear. This can be done by looking through the starter motor opening in the flywheel housing while someone slowly cranks the engine by hand. If the teeth have been badly damaged a new ring gear should be installed. If in fair condition, the teeth can be cleaned up with a file.

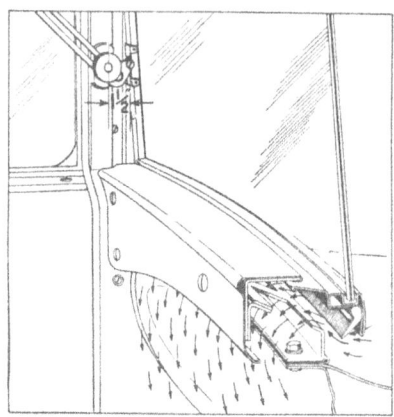

Fig. 581

FRONT COMPARTMENT VENTILATION

To secure maximum ventilation in lower part of front compartment the windshield should be opened 1½" measured from the knurled nut on the swing arm bracket to the slotted screw. (See Fig. 581.) Owners should be instructed regarding this ventilation feature.

BRAKE PEDAL TO CROSS SHAFT ROD

Should a squeak develop due to the brake equalizer beam levers coming in contact with the stops in the rear face of the center cross member, it can be easily overcome by installing the present-design brake pedal to cross shaft rod. This rod being slightly longer than the old-style rod prevents the brake equalizer beam levers from rubbing against the stops.

The new brake pedal to cross shaft rod can be easily distinguished from the old rod as it is slightly longer, the distance from the end of the slot to the end of the rod measuring ⅞" (see Fig. 582) while on the old rod this measurement is approximately ¹¹⁄₁₆".

Fig. 582

Ford Service Bulletin *for October*

Model A Starter Drives

Fig. 584—Starter Drive Parts (Standard) Used on Cars after 492511

Part No.	PART NAME	Price	Part No.	PART NAME	Price
A-11350-C	Starter Drive Assy.	$4.25	A-11382-A	Drive Shaft Spring Screw	$0.10
A-11354-A	Drive Shaft and Pinion Assy.	3.25	A-11377-C	Drive Head Spring Screw	.10
A-11375-C	Starter Drive Spring	.50	A-11379-C	Spring Screw Lock Washer (T-1782)	.01
A-11381-C	Starter Drive Head	.40	A-11383	Starter Drive Spring Clip	.07

Fig. 585—Starter Drive Parts (Special) Used on Cars up to 492511

Part No.	PART NAME	Price	Part No.	PART NAME	Price
A-11350-DR	Starter Drive Assy.	$4.25	A-11382-BR	Drive Shaft Spring Screw	$0.07
A-11354-BR	Drive Shaft and Pinion Assy.	3.25	A-11377-DR	Drive Head Spring Screw	.10
A-11375-DR	Starter Drive Spring	.40	A-11379-C	Spring Screw Lock Washer (T-1782)	.01
A-11381-DR	Starter Drive Head	.40	A-11380-R	Shaft Spring Lock Ring	.01

ANTI-FREEZE SOLUTION FOR COOLING SYSTEM

While there are a number of anti-freezing solutions on the market, probably denatured alcohol and water is the most extensively used.

Below is given the proportion of alcohol and water for freezing temperatures:

Before pouring the solution into the radiator be sure there are no water leaks. Tighten hose connections and inspect water pump packing. Drain off old water and flush radiator out thoroughly.

It must be borne in mind that losses through boiling or evaporation of the alcohol weakens the solution. Consequently to keep the solution at its proper strength, it will be necessary to occasionally add alcohol until the desired hydrometer reading of the specific gravity of the solution is obtained. A hydrometer for this purpose can be purchased from any local accessory store and it is a good plan to obtain one rather than depend on guess-work.

As alcohol is a solvent of pyroxylin, extreme care must be used not to spill any of the solution on the hood.

Capacity Model A Cooling System	10° F. ABOVE ZERO		0° F. ZERO		10° F. BELOW ZERO		20° F. BELOW ZERO	
	Pints Water	Pints Alcohol	Pints Water	Pints Alcohol	Pints Water	Pints Alcohol	Pints Water	Pints Alcohol
3 Gals. (24 Pints)	17	7	15	9	14	10	12	12
Specific Gravity of Mixture	0.9691		0.9592		0.9486		0.9345	

Ford Service Bulletin *for October*

Model A Engine Lubrication

FOR correct engine lubrication a high grade, well refined oil is absolutely essential.

As a guide to the proper viscosity or body of oil for summer and winter conditions, which vary for different territories, the lubrication charts of reputable oil companies should be consulted. In general, an oil having the body of S. A. E. viscosity No. 40 corresponds approximately to our M-515-A and B and is accordingly recommended for summer use.

For winter use, S. A. E. oil specifications No. 20 can be used. It is necessary, however, for winter use, that such an oil have a low cold set.

A large number of oil companies are stamping containers and indicate on their charts S. A. E. numbers. This practice is desirable because of the fact that it assists the owner to obtain the proper oil for his engine.

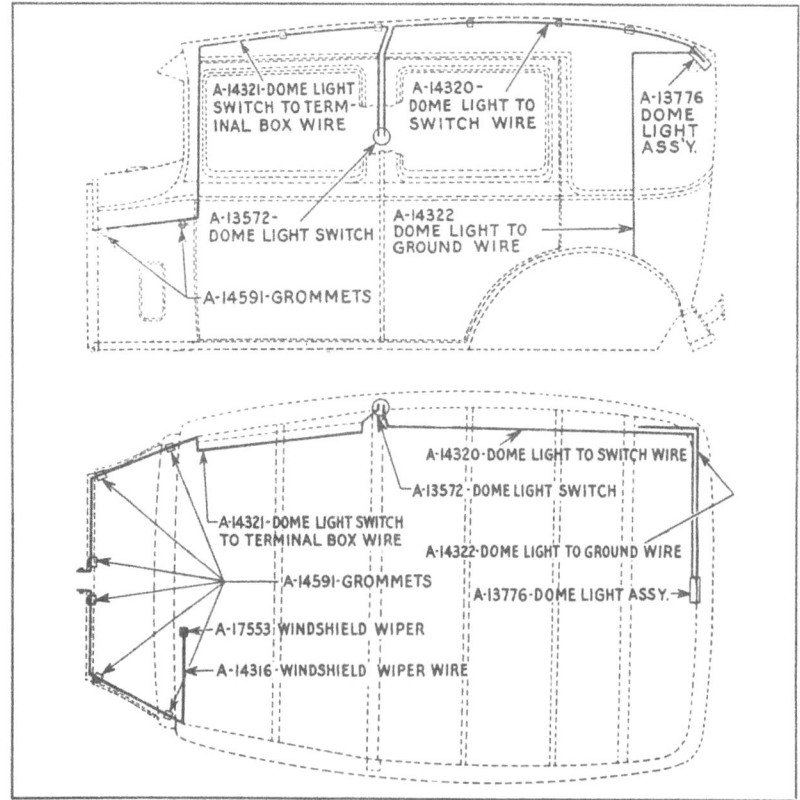

Fig. 593—Model A Fordor Wiring Diagram

Ford Service Bulletin *for November*

NEW FLEXIBLE FRONT END SUPPORT

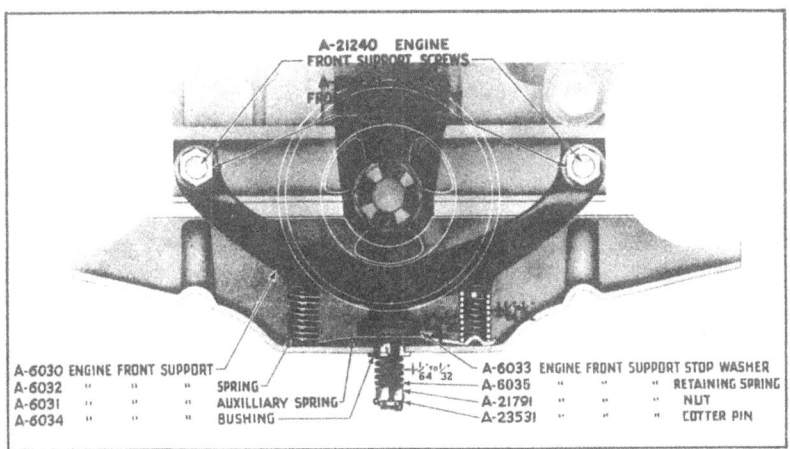

Fig. 597

One of the problems which has constantly confronted automobile engineers is to prevent engine vibration being transmitted to the chassis.

The latest Model "A" improvement along this line is the introduction of a flexible front end support for the engine. The upper half of this support is bolted to the cylinder front cover. The lower half of the support is supported on the cross member by flexible springs (see Fig. 597). These springs allow the engine a free vertical motion, yet hold it within definite limits. The result is an engine support that is simple in design and operation and frees the car from unpleasant vibration periods.

The support is standard on all new cars and trucks and can be easily installed on cars not so equipped by proceeding as follows:

The list price of the support complete with parts is $1.65. The labor charge to the customer for installation must not exceed $6.00.

INSTRUCTIONS FOR INSTALLING

1. Drain water from radiator.
2. Remove hood.
3. Take out mat and floor boards.
4. Disconnect either the battery cable or ground connector from battery.
5. Remove the two accelerator bracket cap screws.
6. Unhook accelerator to carburetor rod, also throttle control rod and lift off accelerator bracket.
7. *Remove the two bolts from both engine rear supports.*
8. Remove both engine pans.
9. Disconnect starter switch push rod and slide rod back out of way.
10. Remove front splash shield.
11. Disconnect cut-out, horn and headlamp wires and remove radiator.
12. Disconnect carburetor adjusting rod from carburetor and loosen the two exhaust pipe bolts.

Fig. 598

13. Remove the two cylinder front cover screws shown in Fig. 598.
14. Remove one of the front spring clips, loosen the other and remove starting crank bearing.
15. Screw off starting crank ratchet nut.
16. Jack up engine sufficiently high to permit withdrawing fan pulley over top of front cross member (To prevent damaging

oil pan, place a small board between top of jack and oil pan.)

17. Saw off cross member at dotted line shown in Fig. 598.

18. Enlarge the ½ inch hole in front cross member to ¾ inch. *Important*: Before enlarg-

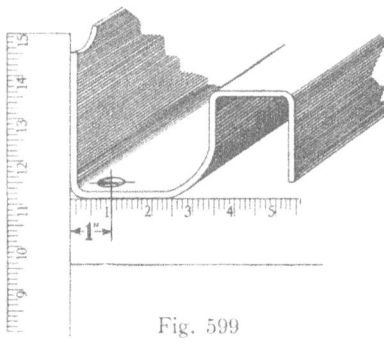

Fig. 599

ing the ½ inch hole make certain that the center of the hole is exactly 1 inch ahead of the rear flange of the front cross member (see Fig. 599) if not, hole must be filed to suit. After enlarging the hole to ¾ inch be sure to remove all burrs and cuttings and thoroughly clean out cross member. *This is important.*

19. Slip leather washer A-6033 over threaded end of support stud, sliding washer back on stud until flat side of washer rests against flat side of support.

20. Place a little grease on the bottom of spring A-6031 and position it in bottom of cross member, lining up hole in spring with hole in cross member. The side of the spring which has the raised spring retainers is placed next to cylinder front cover.

21. Bolt support to cylinder front cover by means of two lock washers A-22330 and two cap screws A-21240.

22. Replace fan pulley.

23. Lower engine sufficiently so that springs A-6032 can be slipped over the two bosses on the support and the raised retainers on the flat spring.

24. Next lower front end of engine down as far as it will go, making sure that neck of leather washer A-6033 lines up with hole in flat spring.

25. Slip brass bushing A-6034 over threaded end of motor support stud which extends through cross member, making sure that neck of bushing enters hole in cross member.

26. Place spring A-6035 over stud, then screw on castle nut, running the nut down sufficiently far to permit locking it in place with cotter key.

27. Next replace the four rear motor sup-

port screws. If it is necessary to jack up the engine, in order to line up the cap screw holes in the rear engine supports, apply the jack under the front radius rod ball cap. After installing the rear motor support screws check the new motor support to make certain that the parts are in exact relationship to each other as shown in Fig. 597. This is important. If the leather washer A-6033 is in contact with flat spring A-6031, or if coil springs are closed, one or both of these conditions will absolutely defeat the action of the support.

28. After tightening the two exhaust pipe clamp bolts, the relation of the front motor support parts should again be checked. This completes the installation of the support and the reassembling of the major parts. The balance of the job can now be built up in the regular way.

29. Before installing the engine pans, bend up the end of both pans as shown in Fig. 600, making sure that the edge of the pan which has been bent up clears front cross member.

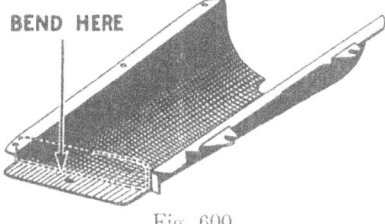

Fig. 600

30. After building up the job make certain there is at least 1/16 inch clearance between the arm on the universal joint housing cap and the center cross member (see Fig. 601). If there is not at least 1/16 inch clearance, it will be necessary to remove the cap and grind off a little stock from the arm until required clearance is obtained.

Fig. 601

FORD SERVICE BULLETIN *for November*

REMARKS

If when the rear motor support arms are bolted to the engine the front support does not have from $\frac{1}{64}$ to $\frac{3}{64}$ clearance between leather washer and top of flat spring, loosen the four bolts holding rear support bracket to motor, also loosen the six small bolts which clamp rear support brackets to frame. Next remove nut and spring and washer on lower end of support stud and jack up front end of motor approximately $\frac{3}{8}$ inch. With the front end of the motor raised $\frac{3}{8}$ inch, tighten the six small bolts which clamp rear support brackets to frame, then tighten the four bolts holding rear support brackets to engine, next lower front end of engine and reassemble spring, washer and nut. This should provide proper clearance between leather washer and flat spring.

If the lower coil spring is closed and there is excessive clearance between leather washer and top of flat spring, repeat the above operation with the exception that the rear end of the engine is jacked up $\frac{3}{8}$ of an inch instead of the front end.

SINGLE PLATE CLUTCH

The new single plate clutch now standard for Model "A" cars and "AA" trucks is composed of two major units, namely, the cover plate assembly A-7563 and the clutch disc assembly A-7550 or AA-7550.

The cover plate assembly consists of a cast iron outer driving plate and a stamped cover plate in which are mounted twelve pressure springs and six release levers. These springs are in direct action against the pressure plate and automatically compensate for all wear of the friction facings. This feature eliminates any necessity of adjusting the release levers.

The driven member or clutch disc assembly is composed of a flat steel disc and two friction facings. The facings are riveted to both sides of the driven disc. The disc is slightly dished in the form of a cone. With this construction the outer and inner edges of the clutch disc facing, start to engage first and as the clutch engages when the pedal is allowed to come back the spring pressure in the clutch flattens out the clutch disc and the entire lining surface picks up the load evenly. This feature assures exceptionally smooth clutch engagement.

Moulded friction facings are used because of their long wearing qualities. They also successfully withstand higher temperatures as they contain no cotton element.

REPAIR PARTS AND EXCHANGES

The only repair parts dealers will stock are the clutch disc assemblies A and AA-7550, clutch disc facings A-7549B and AA-7549, facing rivets A-22993 and pressure plate and cover assembly A-7563.

Should any part of the pressure plate and cover assembly fail, return the entire assembly to the Branch, and a new assembly will be furnished at an exchange price of $3.25 net. The price to the customer will be $4.25.

Under no circumstances will dealers attempt to replace any parts in the pressure plate and cover assembly, as the lever height when under spring pressure must be set with specially designed fixtures.

The following is a list of single plate clutch and related parts which will be supplied through service.

PART NO.	PART NAME	PRICE
A-2455-B	Brake pedal	$1.75
	Brake pedal bushing, A-7508-B	.15
A-6375-B	Flywheel assembly	7.75
A-7006-B	Transmission case	5.00
A-7017-B	Transmission main drive gear	4.00
A-7050	Transmission main drive gear bearing retainer	1.30
	Bearing retainer to transmission case screw washer, A-22218. 75C ea.	.01
	Bearing retainer to transmission case screw washer, A-20718. 75C ea.	.01
A-7501-B	Clutch housing assembly	7.00
A-7506-B	Clutch and brake pedal shaft	.35
A-7507-B	Clutch and brake pedal shaft collar	.45
	Clutch and brake pedal shaft collar pin, A-23789	.01
A-7510-B	Clutch release shaft	.75
A-7511-B	Clutch release shaft arm	.95
A-7515-B	Clutch release shaft fork	1.00
A-7518-B	Clutch housing hand hole cover	.10
A-7520-B	Clutch pedal	1.75
	Clutch pedal bushing, A-7508-B	.15
A-7521-B	Clutch pedal to release arm rod	.30
A-7549-B	Clutch disc facing	.30
	Clutch facing rivet, A-22993. 35C ea.	.01
A-7550	Clutch disc assembly.	3.50
A-7561-B	Clutch release bearing hub	1.95
A-7562	Clutch release bearing spring	.04
A-7563	Clutch pressure plate and cover assembly	7.00
	Clutch cover to flywheel screw, A-20718. 75C ea.	.01
	Clutch cover to flywheel washer, A-22218. 75C ea.	.01
A-7580-B	Clutch release bearing assembly	1.95
A-7600-A2	Clutch pilot bearing	1.25
A-7609-B	Flywheel dowel retainer	.10
A-7620	Clutch housing gasket	.01
A-35121	Floor weather pad upper assembly	.20
A-35123	Floor weather pad lower assembly	.15
	Floor weather pad to No. 1 floor board screw. A-22633	.01
	Floor weather pad to No. 1 floor board screw washer. A-22143. 35C ea.	.01

SPECIAL TRUCK PARTS

AA-7549	Clutch disc facing	.35
AA-7550	Clutch disc assembly	3.75

SERVICE SUGGESTIONS

The pressure springs automatically compensate for all wear of the friction facings. *Readjustment of the release levers must never be made under any circumstances.*

The only adjustment for clutch wear is

Ford Service Bulletin *for November*

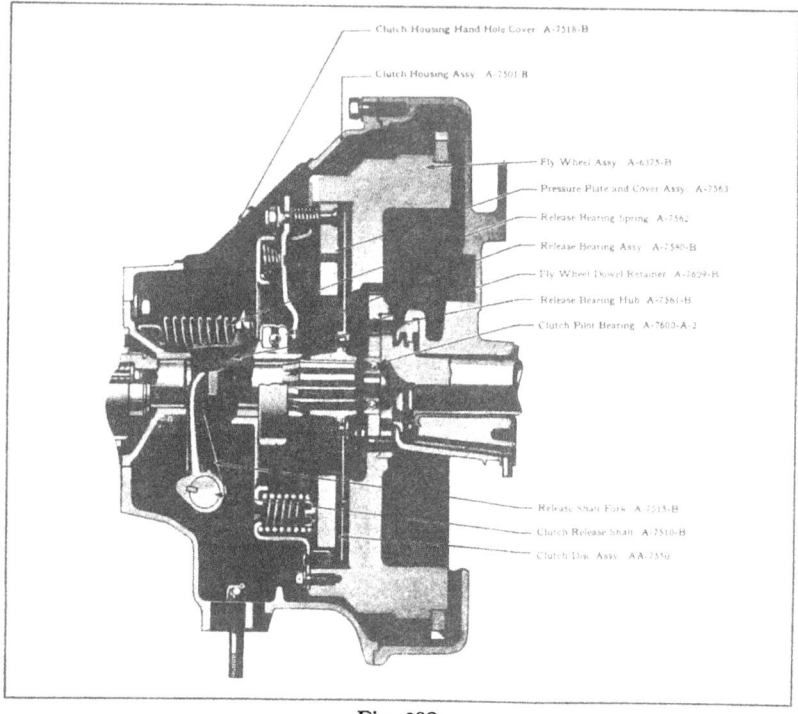

Fig. 602

made at the bottom of the clutch pedal. The pedal *must have 1 inch free-play* or movement before it starts to disengage the clutch.

Grease on friction facings will cause the clutch to chatter during engagement or sometimes slip at high speeds. The remedy is to remove the clutch and install a new set of clutch discs facings.

Occasionally, due to an improperly adjusted clutch pedal or continuous abusive slippage, the clutch pressure plate may develop small radial heat cracks. If the pressure plate is not grooved, and these cracks are not large, simply polish the face and replace the unit. These heat checks will cause no harm.

It is not necessary to return clutch driven members to the factory for replacement of worn friction facings, as this operation may be done in your own shop. Before facings are replaced, make certain that the driven member otherwise is in good condition.

The saw steel driven disc is slightly dished to provide smooth clutch engagement. When new friction facings are installed be certain that the rivets are drawn down tightly.

Drivers should be instructed that riding the clutch pedal is a bad habit, as it causes the clutch to slip. The foot should be placed upon the clutch pedal for a definite purpose only—that is to change gears.

NEW SOLID BRAKE CROSS SHAFT

A new service brake cross shaft assembly has been designed and is now standard on all cars and trucks. The new shaft replaces the old style cross shaft and equalizer assembly, as the old assembly is obsolete and will not be carried for service.

The new shaft is unusually efficient, simple in design and reduces the number of cross shaft parts.

The following parts are required to install the new service brake shaft assembly for car and truck:

FORD SERVICE BULLETIN *for November*

CAR

With hand brake lever in center			With hand brake lever at side	
A-2465-B	$0.25	Brake pedal to cross shaft rod	AA-2465-R	$0.25
A-2466	.20	Brake cross shaft to axle brake rod eye adj.	A-2466	.20
A-2478-B	.10	Brake cross shaft to frame bracket	A-2478-B	.10
A-2479-B	.10	Brake cross shaft to frame bracket shim	A-2479-B	.10
A-2485-D	6.85	Brake cross shaft assembly	A-2845-ER	7.50
A-2499-B	.35	Brake cross shaft to axle brake rod	A-2499-B	.35
		Brake cross shaft guide	A-2491-R	.10
		Brake cross shaft guide bolt 15X	A-20954	.02
		Brake cross shaft guide nut—15 x 1.25C	A-21745	.02
		Brake cross shaft guide cotter—gr. 20, M .50	A-23531 dz.	.03

TRUCK

With hand brake lever in center			With hand brake lever at side	
A-2465-B	$0.25	Brake pedal to cross shaft rod	AA-2465-R	$0.25
A-2466	.20	Brake cross shaft to axle eye adj.	A-2466	.20
A-2477	.20	Brake cross shaft bushing	A-2477	.20
A-2478-B	.10	Brake cross shaft frame bracket	A-2478-B	.10
A-2479-B	.10	Brake cross shaft frame bracket shim	A-2479-B	.10
AA-2485-D	5.35	Brake cross shaft assembly	AA-2485-CR	6.00
AA-2496-C	.80	Brake cross shaft end lever—R. H.	AA-2496-C	.80
AA-2497-C	.80	Brake cross shaft end lever—L. H.	AA-2497-C	.80
A-23383	.02	Brake cross shaft end lever rivet	A-23383	.02
A-2499-B	.35	Brake cross shaft to axle brake rod	A-2499-B	.35
		Brake cross shaft guide	AA-2491-R	.10
		Brake cross shaft guide bolt	A-20747	.02
		Brake cross shaft guide nut—15 x 1.25C	A-21701	.02
		Brake cross shaft guide cotter—gr. 20, M .50	A-23516 dz.	.03

The following cross shaft and equalizer parts are obsolete and will not be supplied after present stocks are exhausted. If any of these parts are ordered, the parts necessary to install the new brake cross shaft should be furnished at the price of the part or parts so desired.

A-2470	Equalizer operating shaft and bushing assembly—L. H.
A-2480	Equalizer beam
AA-2481	Equalizer beam lever
AA-2485-AR, B&C	Cross-shaft assembly—R. H.
AA-2486-AR, B&C	Cross-shaft assembly—L. H.
AA-2494-AR	Cross-shaft—R. H.
AA-2495-AR	Cross-shaft—L. H.
AA-2496-AR	Cross-shaft end lever—R. H.
AA-2497-AR	Cross-shaft end lever—L. H.
A-4520	Universal joint housing cap outer assembly

When the solid brake cross shaft is installed on cars equipped with adjustable brake rods, it will be necessary to substitute A-2466 adjustable eye for the A-2461 clevis on the rear rods.

When installing solid brake cross shaft on cars and trucks equipped with hand brake lever at side, an A-2491-R or AA-2491-R cross shaft guide must be installed (see Figs. 603 and 604).

The brake rod holes in cross member must also be enlarged to provide sufficient clearance for the brake rods (see Figs. 603 and 604).

When used on cars equipped with solid brake rods, it will be necessary to change the solid rods into adjustable rods. This can be done by sawing off the ends of all four rods 33" from the center of the button on the rods (see Fig. 605) and running a $\frac{5}{16}$—24 thread back 2" from the end of the rod, then installing locknut A-21700 and adjustable eye A-2466. The length of the rod measured from the centerline of both eyes should be adjusted to not more than $51\frac{7}{16}$ or less than $51\frac{1}{2}$ (see Fig. 605).

When installing a solid brake cross shaft on trucks equipped with adjustable front brake rods, it will be necessary to shorten the in-

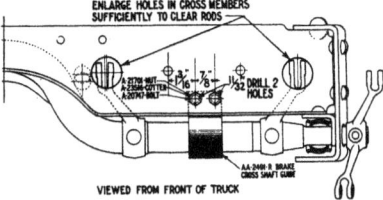

Fig. 604

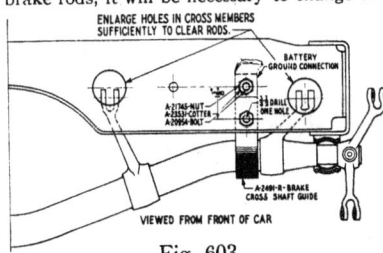

Fig. 603

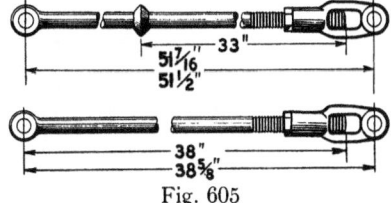

Fig. 605

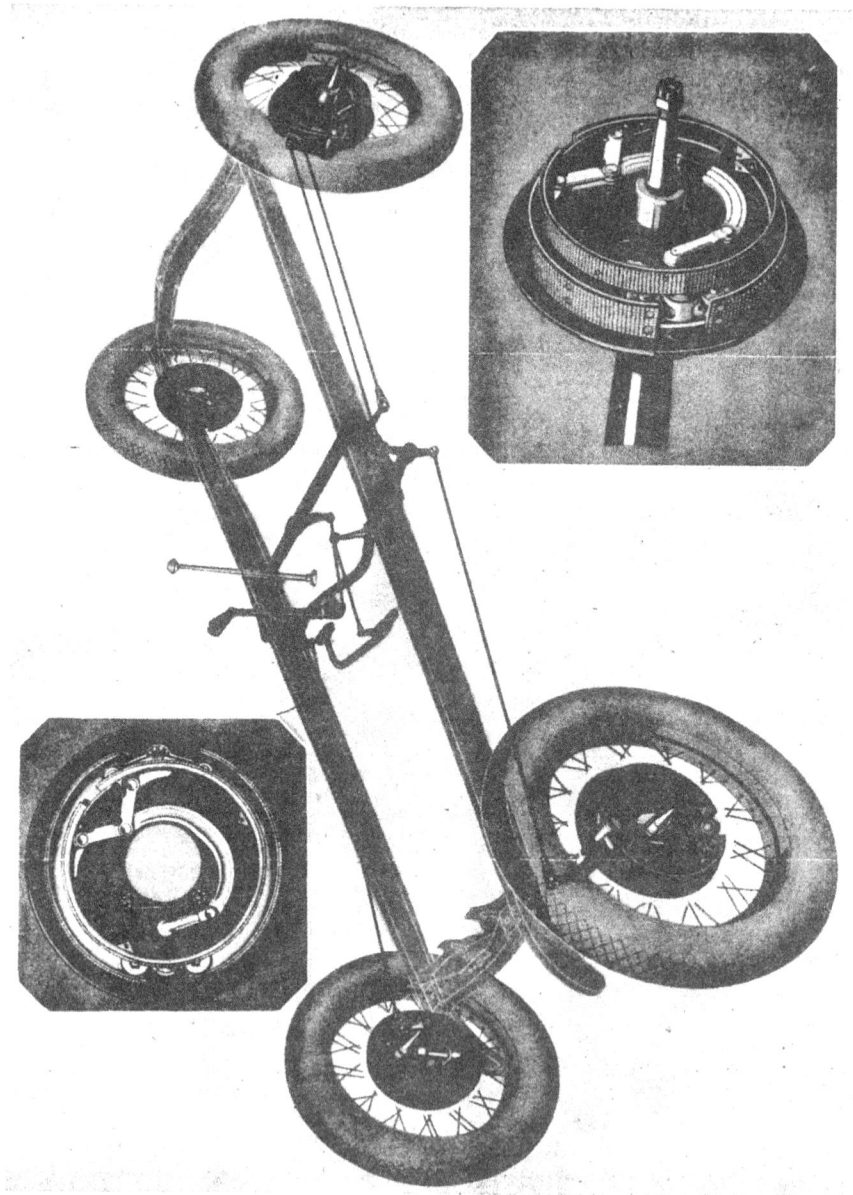

Fig. 607—Model "A" Braking System with New Solid Brake Cross Shaft

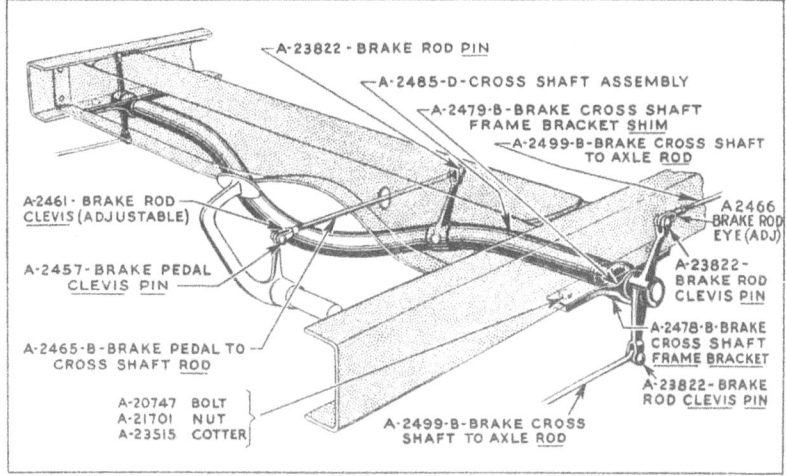

Fig. 606

termediate rod AA-2500. This is done by sawing ⅝" off either end of rod, threading the rod 2" back with a ₇⁄₁₆—24 die and installing lock nut A-21700 and adjustable eye A-2466. The length of the rod measured from the centerline of both eyes should then be adjusted so that it measures not less than 38⅝ or more than 38¹¹⁄₁₆ (see Fig. 605).

When installing the new cross shaft on trucks equipped with solid front brake rods, it will be necessary to make an adjustable rod out of the front rods in the same manner as described for the car. It will also be necessary to rework the intermediate rod AA-2500 as described in preceding paragraph.

INSTALLATION AND ADJUSTMENT

The oilless type bearings on the ends of the new solid brake cross shaft are assembled to the frame by the use of A-2479-B shims and A-2478-B brackets (see Fig. 606). When assembling, a small amount of grease is required in bracket as the outside of the bearing must rotate slightly inside of the bracket in order to preserve alignment when flexure of the frame occurs.

After installing cross shaft install brake pedal to cross shaft rod. The adjustable clevis end of the rod is assembled to the brake pedal—the non-adjustable end, to the lever on the cross shaft (see Fig. 606).

A slightly different procedure is followed in adjusting this rod depending on whether the car is equipped with a single plate clutch or a multiple disc clutch as the brake pedals used on cars equipped with the single plate clutch are provided with a stop.

If the car is equipped with a multiple disc clutch adjust the rod as follows: Hold the tip of the rod against the rear flange of the center cross member, then adjust the clevis on the opposite end of the rod until the brake pedal arm clears the underside of No. 1 floor board by ½" to ¾".

If the car is equipped with a single plate clutch, pull the brake pedal all the way back until it is against its stop—then adjust the rod until there is approximately ₁⁄₁₆" clearance between end of rod and rear flange of center cross member (leaving a little clearance between end of rod and cross member prevents any possibility of the end of the rod rubbing against the cross member and causing a squeak).

After adjusting the brake pedal to cross shaft rod, assemble side pull rods to brake operating and cross shaft end levers. When assembling the side pull rods, pull the brake operating levers on the front axle *back* and the brake operating levers on the rear axle *forward* (taking up all idle movement). Then with the levers in this position adjust the length of the side pull rods so they can be assembled to brake operating and cross shaft end levers.

Next adjust brakes by turning up the adjusting wedges as described on page 202 of the January, 1928, Bulletin.

After the brake rods have been correctly adjusted car owners should be notified that **this adjustment must not be altered.** Service brake adjustments must be made only by means of the adjusting wedge at each brake.

Ford Service Bulletin *for February*

NEW STEERING GEAR

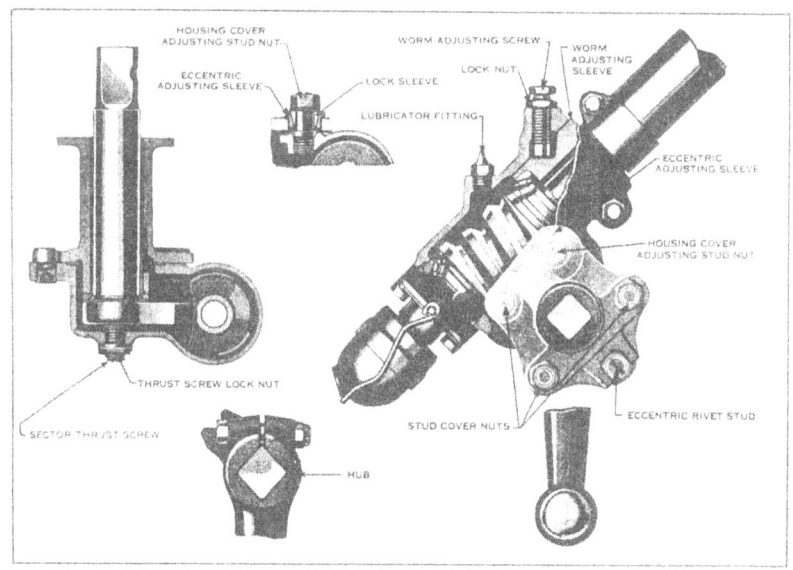

Fig. 637

A percentage of Model "A" cars are now equipped with a new design steering gear (see Fig. 637). The new steering gear is an exceptionally efficient design. It responds easily to movements of the wheel under the hands of the driver, is especially geared for handling balloon tires, and due to its efficient construction makes for unusually easy operation without possibility of the wheel being jerked from the driver's hands by ruts or bumps in the road.

The thrust on the worm is taken up by two roller thrust bearings, placed at each end of the worm. This insures proper alignment and prevents any binding of the steering worm shaft.

The new gear is known as the hour glass worm and two tooth sector type. The worm is so cut that the sector teeth have no play or lash in the center (the straight ahead driving position) but with gradually increasing lash toward ends. This provides against binding at extremes after adjustment for normal wear.

The steering column is clamped to the bearing adjusting sleeve, which permits re-servicing in part and gives ample strength and proper alignment.

LITTLE ATTENTION REQUIRED FROM A REPAIR STANDPOINT

Owing to efficient design and sturdy construction, the Model "A" steering gear assembly with ordinary care will last indefinitely and should require little attention from a repair standpoint. In time, of course, it will require adjustments to compensate for natural wear.

ADJUSTMENTS

When making adjustments, the front wheels of the car should be jacked up and the drag link disconnected from steering arm in order to effect a satisfactory adjustment.

There are three adjustments which can be made with the steering gear assembled in the car, namely: End play in worm sector; end play in steering shaft; proper mesh of sector teeth in worm. When it is necessary to make any one of these adjustments, the other two adjustments should also be checked. Adjustments should always be checked in the following order:

ADJUSTMENT OF END PLAY IN WORM SECTOR

First see that housing cover nuts (see Fig. 638) are securely tightened. Next turn steering

FORD SERVICE BULLETIN *for February*

Fig. 638

wheel to either extreme, then back one-eighth of a turn. Gripping steering arm at hub (see Fig. 639), the shaft should move freely

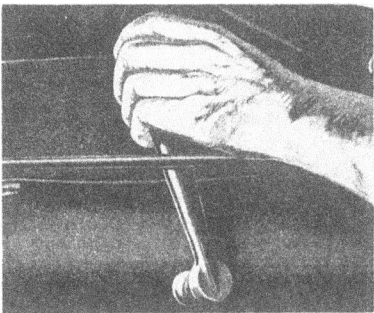

Fig. 639

when turned back and forth, without a particle of end play. Adjust as required by means of sector thrust screw at side of housing next to motor (see Fig. 640). A special offset screw driver is required for this purpose. (K. R. Wilson can supply this tool.) After

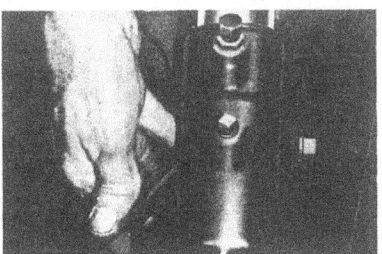

Fig. 640

making adjustment, be sure to tighten lock nut (see Fig. 637), then reinspect for end play and freedom.

ADJUSTMENT FOR END PLAY IN STEERING SHAFT

To adjust for end play in steering shaft or between worm and roller bearing thrusts, turn steering wheel to either end stop, then back up one-eighth of a turn, or to a point where there is lash of steering arm. This

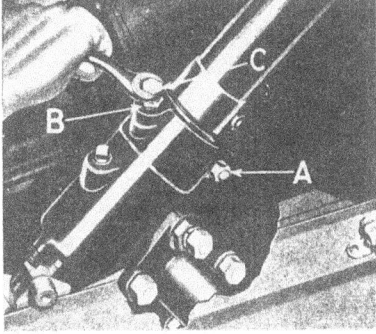

Fig. 641

leaves the steering shaft bearings free of side thrust. Next loosen housing clamp bolt (see "A," Fig. 641), and lock nut "B," on worm adjusting screw "C." Turn down adjustment screw tightly with a six-inch wrench,

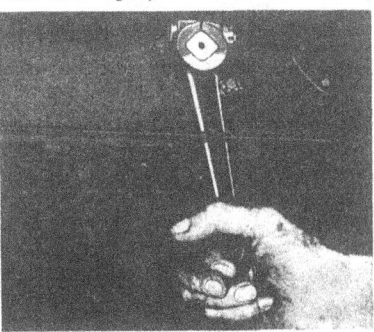

Fig. 642

then back off one-sixth of a turn. Next tighten lock nut and housing clamp bolt securely. Turn steering wheel from extreme to extreme positions and test for stiffness.

ADJUSTMENT FOR PROPER MESH OF SECTOR TEETH IN WORM

Turn steering wheel to the mid-position

FORD SERVICE BULLETIN *for February*

of its complete travel or turning limits. (Drag link previously disconnected.) Shake steering arm to determine amount of lost

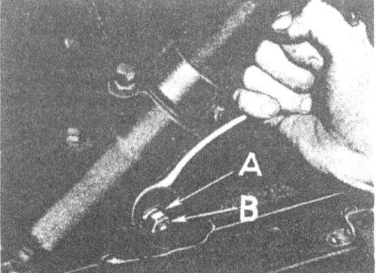

Fig. 643

motion (see Fig. 642). Next loosen the three housing cover stud nuts (Fig. 638) exactly one-quarter turn, then loosen housing cover adjusting stud nut (see "B," Fig. 643) one-half turn. Turn the eccentric adjusting sleeve "A," clockwise, very gradually, checking at each movement the amount of lost motion still existing at the steering arm. Adjust only sufficiently tight to eliminate all lash of steering arm (no more), being sure to finish movement of eccentric adjustment sleeve "A" in clockwise direction. Turn steering wheel throughout full travel to test for free operation. If too tight, turn eccentric adjusting sleeve "A" counter-clockwise to free and readjust, as above, more carefully. Next securely tighten housing cover adjusting stud nut (see "B," Fig. 643) and follow by tightening housing cover nuts (see Fig. 638). It is important that the adjusting stud nut be tightened before tightening housing cover nuts.

CENTRALIZATION OF TOOTH CONTACT

The foregoing adjustments will suffice in practically every instance. Occasionally, however, even after all three of these adjustments have been carefully made and checked, there may still be an unequal amount of lash between sector teeth and worm at points equi-distant from central position of worm. To compensate for any unequalities in lash, an eccentric rivet adjustment is provided (see "M," Fig. 644). By means of this adjustment, the sector shaft can be shifted to either side of the worm centerline. To make this adjustment, however, it is necessary to remove the steering gear assembly from the car.

The worm has a left-hand thread, consequently, turning the steering shaft to the right, moves the sector teeth to the lower end of the worm (see "B," Fig. 644). Turning the steering shaft to the left moves the sector teeth to the upper end of the worm (see "A," Fig. 644).

In making this adjustment, the check must start with the sector shaft teeth meshed at the center of the worm (see "D", Fig. 644). To test for the center of the sector shaft being on the center line of the worm, turn steering shaft to the left as far as possible against stop, then turn to the right one and one-half turns. From this point, turn the shaft by shortest movement until the wheel keyway (see "A," Fig. 647) lines up with the adjusting screw "B."

Next turn the steering shaft one-half revolution to the right (using the keyway as a marker), and shake the steering arm to note the amount of play or lash at this point.

Then turn the shaft back to the left one complete revolution, or in other words, one-half revolution to the left of center, and shake steering arm to see if there is any difference in the amount of lash in the arm as compared with other location. See Fig. 642.

If there is less lash when steering shaft is turned to the left, slightly move eccentric rivet "M" in a clockwise direction (see "B," Fig. 644).

If the lash is less when the shaft is turned

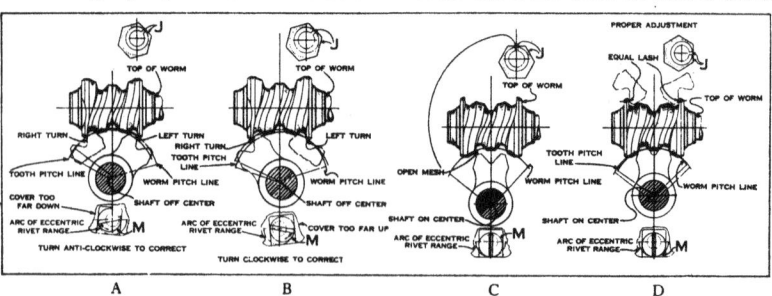

Fig. 644

FORD SERVICE BULLETIN *for February*

to the right one-half turn than it was when the shaft was turned to the left, move the eccentric rivet a small amount in an anti-clockwise direction (see "A," Fig. 644).

When the amount of lash of the steering arm is equal when the steering shaft is turned both right and left one-half turn from central position, adjust for proper mesh of sector teeth in worm as described under heading "Adjustment for Proper Mesh of Sector Teeth in Worm."

After making final adjustment, securely tighten cover adjusting stud nut, then follow by tightening housing cover nuts. It is important that the adjusting cover stud nut be tightened first.

STEERING GEAR OVERHAUL

The simplicity of the new design steering gear permits easy dismantling and the replacement of any parts. To overhaul the new gear proceed as follows:

DISASSEMBLING

Remove steering assembly from car.
Drain out lubricant.
Place steering gear assembly in vise. Loosen steering column clamp bolt nut and withdraw steering column from adjusting sleeve. (See Fig. 645.)

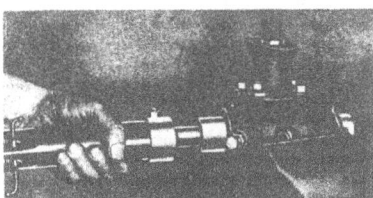

Fig. 645

Remove cover nuts and eccentric sleeve locknut and lift off cover and cork oil seal and withdraw sector shaft (see Fig. 647).

Back off worm adjusting screw all but two threads, loosen housing clamp bolt, then insert a wedge in housing clamp slot "B," and withdraw shaft assembly, as shown in Fig. 646.

Take lower bearing out of housing if it did not come out with worm.

INSPECTION

The parts should now be washed in kerosene and each part carefully examined for wear. The bushings in the housing should also be inspected. New gaskets will usually prove more satisfactory when reassembling, although an old gasket, if in good condition, will render satisfactory service. Inspect each gasket before installing; see that the surface against which it fits is clean and in good condition.

Before assembling, thoroughly lubricate all moving parts. Draw all bolts, nuts and cap screws down tightly, making sure to replace lock washers and cotter pins as required.

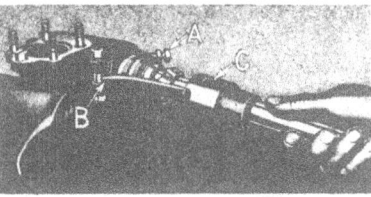

Fig. 646

ASSEMBLING

First grease worm, cones and bearings, then insert steering shaft assembly into housing, as shown in Fig. 646. Be sure lower bearing is in proper position and that worm adjusting screw seat "C" lines up with adjusting screw "A." Next make steering shaft adjustment as described under heading "Adjustment for End Play in Steering Shaft."

After making adjustment, turn steering shaft until keyway in shaft (see "A," Fig. 647) lines up with worm adjusting screw "B." The cover bushings, sector shaft and thrust washer should now be thoroughly lubricated. After lubricating these parts, replace cork gasket, see "A," Fig. 648. Slip thrust washer "B," Fig. 648, over end of sector shaft, making sure that neck of washer points toward housing cover as shown and the slot in the eccentric adjusting sleeve (see "C," Fig. 647) points towards worm.

Next slip housing cover over ends of studs and just start the cover nuts, making sure

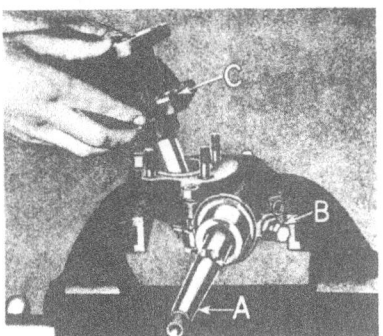

Fig. 647

FORD SERVICE BULLETIN *for February*

Fig. 648

that the lock sleeve (see Fig. 637) is in place. Before tightening cover nuts, back off sector adjusting screw two or three turns (see Fig. 649). After backing off adjusting screw, tighten cover nuts. Next temporarily place steering arm on sector shaft, as shown in Fig. 639, and adjust for end play in worm sector, as described under heading "Adjustment of End Play in Worm Sector." After adjusting sector shaft for end play, adjust for proper mesh of sector teeth in worm, as previously described.

The steering column can now be installed. When installing the column, be sure that the center between control rods lines up with worm adjusting screw. This insures correct location of spark and throttle rods.

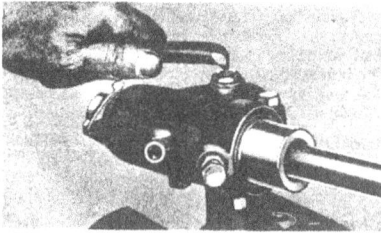

Fig. 649

The assembly can now be installed in car. Be sure the steering gear is filled with gear lubricant. The new design assembly requires approximately 4½ ounces of lubricant. A fluid lubricant of the consistency of 600-W should be used. *Do not use cup grease.*

NEW TWOLITE HEADLAMPS

The A-13004-A and A-13005-A Ford headlamps formerly used on the Model "A" have been superseded by the new design Ford Twolite headlamps which are supplied in two types—a two bulb type A-13005-C for cars without cowl lamps and a single bulb type A-13005-D for cars with cowl lamps. These lamps are nickel plated. Black enamel headlamps listed under part Nos. A-13004-C and D are furnished for trucks and commercial jobs.

The new Twolite headlamps have created an entirely new standard of lighting efficiency. Their design and workmanship is unexcelled by any headlamp regardless of price.

The new lamps are supplied with 6-8-volt double filament double contact gas filled bulbs. Both of the filaments are 21 candle power. The lower filament provides a beam which gives exceptionally brilliant road illumination for high speed driving. The upper filament provides a downward tilting beam which illuminates the entire road surface close to the car. These filaments provide the maximum in lighting efficiency for both city and country driving.

The A-13005-C two bulb headlamp has a separate 2 candle power bulb mounted in the upper part of the reflector for parking purposes.

In addition to the standard 21 candle power Twolite bulbs we are also releasing through service (in those States in which the law permits) a 32-21 candle power bulb A-13007-D. This bulb gives a driving light of greater beam intensity than the standard bulb.

OLD STYLE HEADLAMPS CANNOT BE CONVERTED INTO THE NEW TYPE

Due to the difference of the spacing of the filament in the bulbs, the old design A-13004 and 5-AR and BR headlamps cannot be converted into the new Twolite type by installing new lens and bulbs. The new headlamp bulbs have the filaments equally spaced above

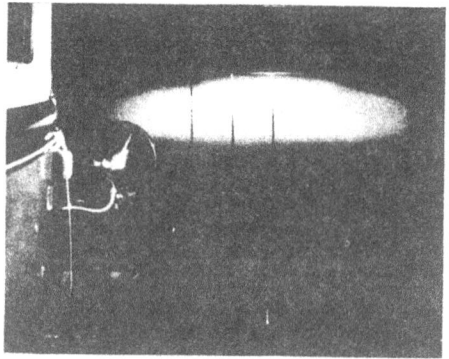

Right Headlamp Properly Focused and Aligned

Fig. 650

FORD SERVICE BULLETIN *for February*

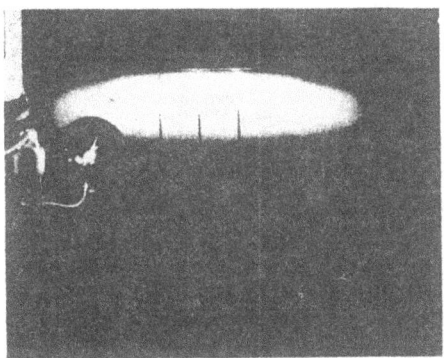

Both Headlamps Properly Focused and Aligned

Fig. 651

and below the central axis of the bulb while on the old style "H" headlamps the major filament was located directly on the central axis of the bulb.

If an owner desires to replace old style headlamps with the new design, it will, in addition to replacing the lamps, be necessary to replace the headlamp wiring and switch. Figs. 653 and 654 show the new wiring diagram.

INSTRUCTIONS FOR FOCUSING AND ALIGNING

Ford Twolite Headlamps with A-13060-B Ford Twolite Headlamp lens and A-13007-C, 21-21 c. p. and A-13466, 2 c. p. bulbs must be properly focused and aligned.

The lamps are correctly adjusted at the factory when they are assembled to the car. Where replacements are required it will of course be necessary for dealers to make adjustments for the owner. This is done as follows:

Align and focus headlamps with empty car standing on a level surface in front of a white wall or screen 25 feet from the front of the headlamps. This wall must be in semi-darkness or sufficiently shielded from direct light so that the light spots upon it from the headlamps can be clearly seen, and must be marked off with black lines as shown in cuts.

Focus

1—It is important that lens be installed in door with the word "top" at top of door and with all lettering reading properly from front.

2—Turn on upper beam.

3—Focus by means of screw at back of lamps, first one lamp and then the other, adjusting the bulb filament at the focal center of the reflector to obtain an elongated elliptical spot of light on the wall, with its long axis horizontal (see Fig. 650). In focusing, adjust the bulb to obtain as good contrast and as well-defined cut-off across the top of the spot of light as possible.

With lamps thus focused for the upper beam the lower beam will be in satisfactory position.

No adjustment is necessary for the small bulb for parking light.

Alignment

1—Headlamps are aligned by moving lamps after nut at bottom of bracket has been slightly loosened.

2—The tops of the bright spots on the 25-foot wall are to be set at a line 37 inches above level of surface on which car stands. (See Fig. 652.) With tops of bright spots thus set for empty car, the headlamps comply, under all conditions of loading, with the requirements of the various states.

3—The beam of light from each headlamp is to extend straight forward; that is, the centers of the elliptical spots of light must be 30 inches apart.

Proper alignment of headlamps is readily checked by means of a horizontal line on the

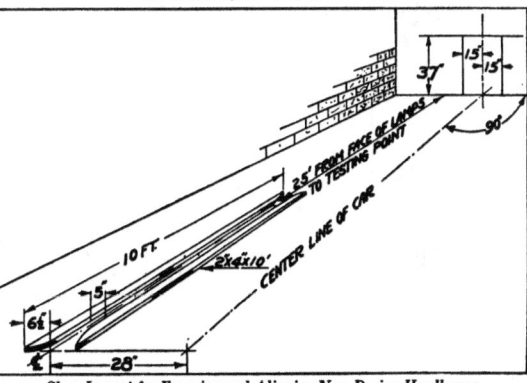

Shop Layout for Focusing and Aligning New Design Headlamps

Fig. 652

FORD SERVICE BULLETIN *for February*

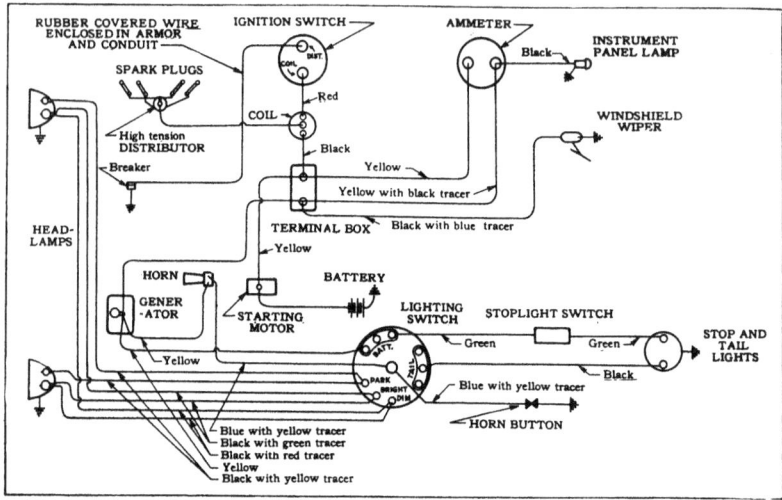

Model "A" Wiring Diagram—*Cars not equipped with Cowl Lights*

Fig. 653

wall in front of the car, 37 inches above the level surface on which car stands, and two vertical lines 30 inches apart, each one 15 inches from center line of car. Proper alignment of car relative to marks on the wall may be readily provided by use of wheel guide blocks for one side of the car, as shown in cut. If it is impossible to tie up the floor space required by these blocks, marks painted on the floor may be used to show where one set of wheels should track and where the car should be stopped.

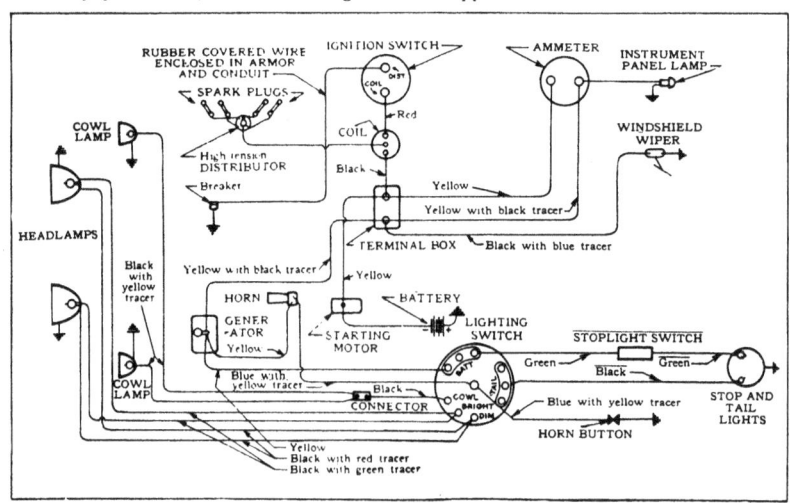

Model "A" Wiring Diagram—*Cars equipped with Cowl Lights*

Fig. 654

FORD SERVICE BULLETIN *for August*

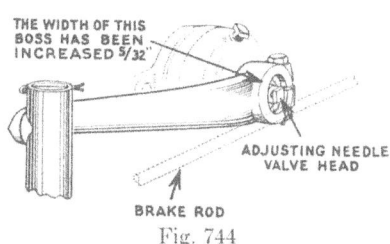

Fig. 744

WIDTH OF FRONT SHOCK ABSORBER ARM BOSS INCREASED

The width of the large boss on the front shock absorber arms has been increased $\frac{5}{32}''$. (See Fig. 744.)

This change prevents any possibility of the adjusting needle valve head extending beyond the boss sufficiently to permit it striking against the brake rods.

The width of the large boss on the rear shock absorber arms was increased sometime ago.

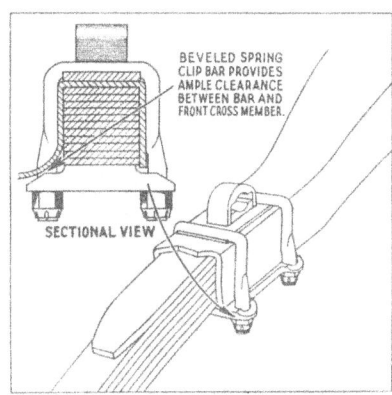

Fig. 745

SPRING CLIP BAR REDESIGNED

To prevent any possibility of the front cross member coming in contact with the spring clip bar, the ends of the spring clip bars are now beveled instead of being machined flat. (See Fig. 745.)

This change provides ample clearance between cross member and spring clip bar at all times.

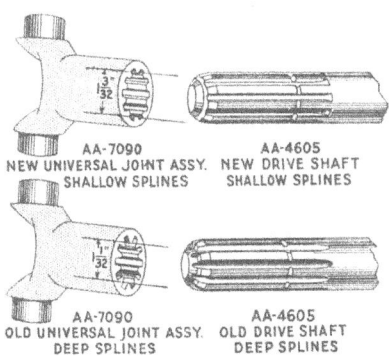

Fig. 746

CHANGE IN DRIVE SHAFT AND UNIVERSAL JOINT SPLINES

The splines on the end of the AA-4605 drive shaft and AA-7090 universal joint have been redesigned.

The new splines are slightly shallower than the old design. This changes the diameter of the splines in these parts. (See Fig. 746.)

When replacing an old drive shaft with one of the new shafts with the shallow splines, it will be necessary to install a new universal joint with the $\frac{1.083}{1.085}$ splines, as the old style universal joint cannot be used with the new style drive shaft. Old style drive shafts can, however, be used with the new universal joint.

In addition to the difference in the width of the splines, it will also be noted that the splines on the old drive shaft are rounded on the bottom, whereas those on the new drive shaft are square.

CLEANING SPORT COUPE TOPS

Several requests have been received for information on how to clean the material on Sport Coupe Tops. After repeated tests our Laboratory advise that a good grade of saddle soap or Ivory soap worked into a lather and applied with a sponge or soft cloth, will satisfactorily clean the majority of these tops. Our M-217 upholstery cleaner can also be used with success. However, care must be exercised when using the cleaner, as too much pressure will remove the grain from the material. This applies to both pyroxylin coated brown and light gray material.

USE "B" TYPE WIPER FOR SERVICE REPLACEMENT

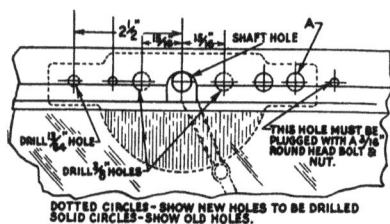

Fig. 658

When dealer's stocks of A-17553-A wipers are exhausted, it will be necessary to use the new design wiper A-17553-B, as the "A" type wiper will no longer be supplied.

The new wiper can be easily installed on old cars by relocating three of the five drilled holes in the windshield frame.

To drill the new holes it is first necessary to remove the windshield.

Five drilled holes are used in mounting the new wiper on the frame. However, as two of the old wiper holes are in the correct location for installing the new wiper it will only be necessary to drill three new holes, namely, one 11/32" and two 3/8" (see dotted circles in Fig 658).

The old hole shown at "A" in sketch which is used in bolting the new wiper to the frame is 11/32" larger than is necessary. No trouble, however will be experienced from this oversize providing the wiper mounting bolts are securely tightened.

When the new wiper is installed one of the holes which were used when mounting the old wiper will remain uncovered. This hole can be plugged with a 3/16" round head bolt and nut.

Wipers are guaranteed for four months. Should the internal mechanism of the wiper fail within that period a new wiper should be furnished free.

No installation charge must be made for replacing an old style wiper with the new design within the new car guarantee period. Beyond this period a recommended labor charge of $1.00 may be made.

The clutch used in the Model "A" car and "AA" truck is a dry disc clutch. Under no circumstances must oil or grease be used in these clutches.

NEW MUFFLER OUTLET PIPE CLAMP

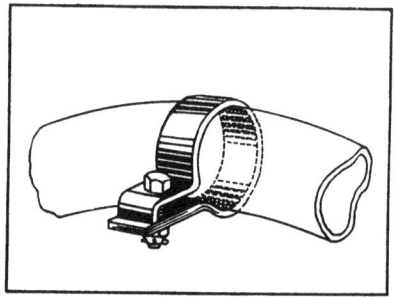

Fig. 659

To hold the muffler outlet pipe more rigidly and insure a tighter clamp on the pipe, muffler outlet pipe clamp (A-5256B) has been superseded by clamp (A-5256C). See Fig. 659.

When replacing an old clamp with the new design it will be necessary to drill a 11/32" hole in frame side member 1 11/32" back of the present hole. See Fig. 660.

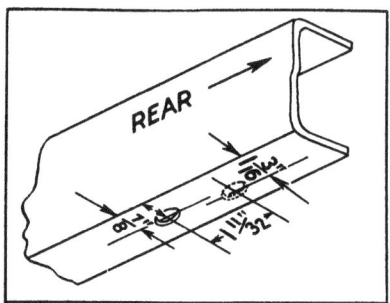

Dotted circle shows location of new hole to be drilled

Fig. 660

DIFFERENTIAL DRIVE GEAR AND DRIVING PINION

To simplify production of gears and pinions we have discontinued manufacture of the A-4209AR, gear and pinion, 10-37 ratio, and will hereafter supply a 9-34 ratio gear and pinion under part number A-4209AR.

When installing sport light on Fordor be careful not to run drill through windshield wiper wire.

Ford Service Bulletin *for March*

USE "AR" WHEELS WITH "AR" HUBS AND "B" WHEELS WITH "B" HUBS

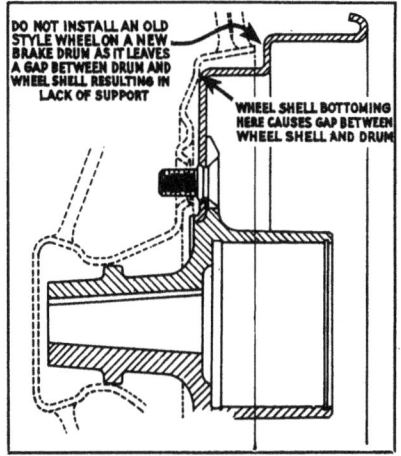

Fig. 661

The A-1015AR wheel is designed for use only with A-1105AR and A-1115AR hub and brake drum. The A-1015B wheel is designed for use with A-1105B and A-1115B hub and brake drum.

Under no circumstances must an "AR" wheel be used with a "B" hub or vice versa. Use of an "AR" wheel with a "B" hub leaves a space between wheel shell and brake drum, thus losing the stiffness that this assembly should have. See Fig. 661.

Use of a "B" wheel with an "AR" hub does not allow sufficient threads on the hub bolts to project through the wheel to permit the wheel nuts to be securely tightened. See Fig. 662.

REAR BRAKE ROCKER ARM REDESIGNED

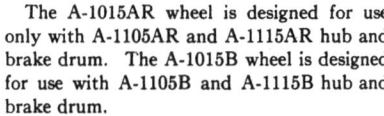

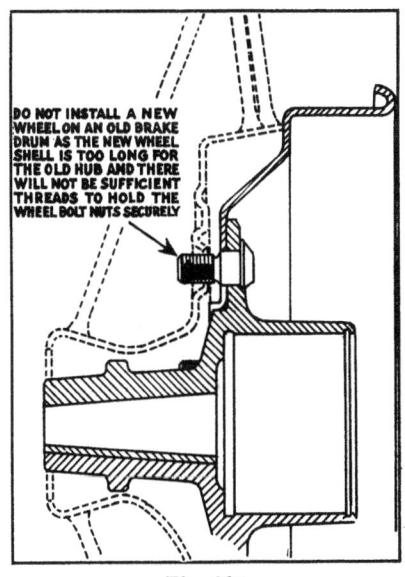

Fig. 662

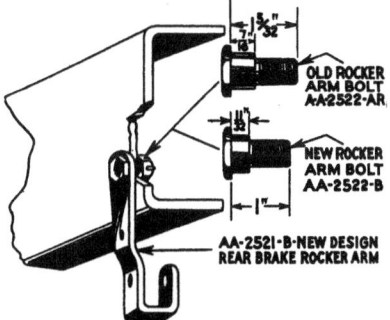

Fig. 663

The rear brake rocker arm AA-2521-B has been redesigned and changed from a forging to a stamping. The change in design alters the length of the bolt used in attaching the arm to the frame. See Fig. 663.

Bolts with the $\frac{5}{32}''$ shoulder must be used only with the old style arm. Bolts with the $\frac{1}{8}''$ shoulder are to be used only with the new rocker arm.

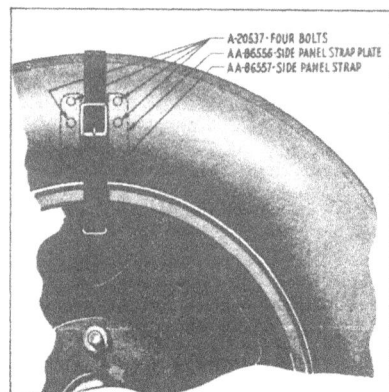

Fig. 747

NEW SIDE PANEL STRAP PLATE

To hold the spare tire more securely against the side of the panel body, the AA-86556 side panel strap plate has been redesigned and an AA-86557 strap added.

The new strap and plate holds the spare tire rigidly against the side of the body.

If you should find it necessary to replace any of the old style parts, the new plate and strap can be easily installed as shown in Figs. 747 and 748.

MOST COIL TROUBLES DUE TO NEGLECT

An examination of coils sent in by dealers as alleged defective material frequently show that the trouble experienced was entirely due to failure on the part of mechanics and owners to keep the bakelite insulator on the coil clean (see Fig. 749). After cleaning and scraping the insulator at the factory the coils invariably check 100%.

When dust and moisture are allowed to accumulate on the insulator, it sets up a path for leakage of the secondary current to ground instead of permitting the current to go on to the distributor and spark plugs.

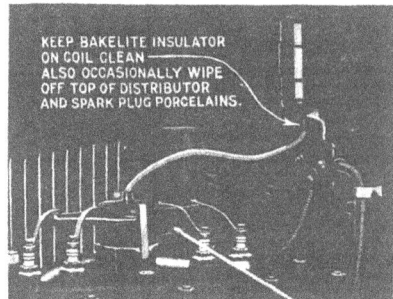

Fig. 749

At first dust and moisture on the insulator may cause only a slight miss in the engine. Eventually it causes the engine to misfire continuously or cut out entirely, thus creating the impression that the coil is dead. As previously stated, this is due to the secondary current passing along the moisture and grounding on the metal case on the coil. In following this course, the current gradually forms a carbon path that resembles a crack in the insulator (see Fig. 750). By cleaning the insulator and scraping off the carbon with a sharp knife the coil will again operate satisfactorily.

Instruct your mechanics, also make certain that owners understand the importance of keeping the coil insulator, also the spark plug porcelains, and the top of the distributor clean. Any foreign matter around the electrical insulation allows leakage to take place and

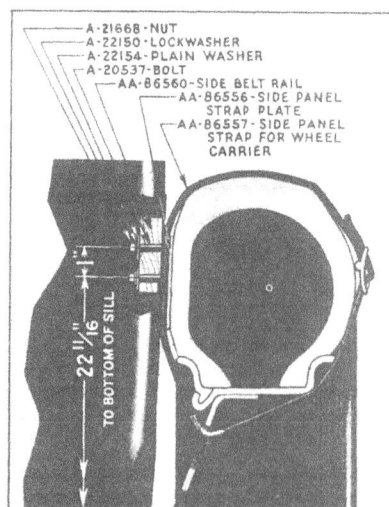

Fig. 748

cuts down the amount of current delivered to the plugs.

Another common cause of coil troubles is the failure of owners when shutting off the engine to push in on the cylinder of the Electrolock sufficiently far to permit it to snap back into the locked position. If the breaker points are closed when the lock is not all the way in, it permits the current to flow through the coil, causing overheating and consequent damage to the coil.

Make certain that every owner understands that when shutting off the engine it is necessary to push the lock all the way in until it snaps back into the locked position.

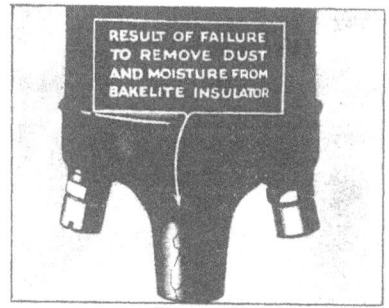

Fig. 750

PACKAGE TRAY INSTALLATION

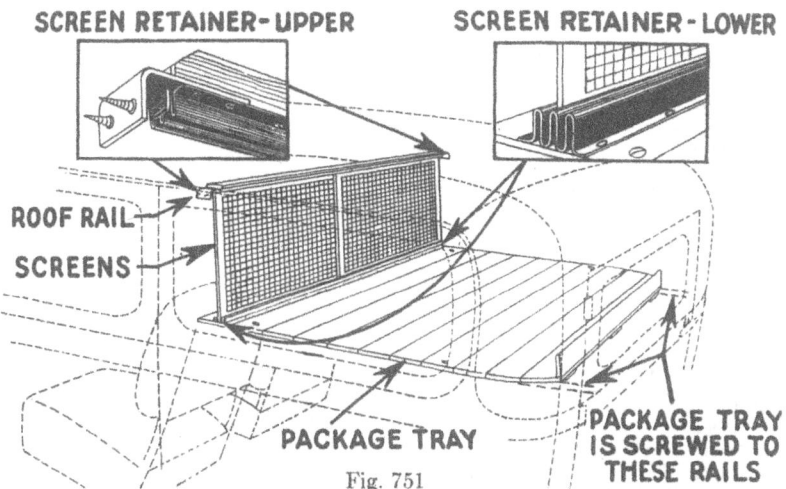

Fig. 751

From letters received, it is evident that all dealers are not entirely clear regarding the proper method of installing the package tray and screen in De Luxe delivery bodies.

To install this assembly first place the package tray in position on the belt rails and screw the tray to both rails. Screw holes are provided in the package tray for this purpose.

Next insert package tray screen into the lower retainer, (the lower retainer is already attached to package tray) then install the upper retainer, screwing both clamps to side roof rail as shown in Fig. 751.

Parts Used in Making the Installation

Part No.	Name	List Price	No. Req'd Per Car
A-132905	Package tray assembly	$6.50	1
A-22675	Package tray assembly screw	.01 ea.	4
A-132966	Package tray screen run top assembly	1.75	1
A-132952	Package tray screen assembly	2.00 ea.	2
A-22627	Screw	.01 ea.	4

The list price of all necessary parts for package tray installation is $10.00. Dealers' discount 25%.

Ford Service Bulletin *for August*

VACUUM AUTOMATIC WINDSHIELD WIPER USED ON TUDORS

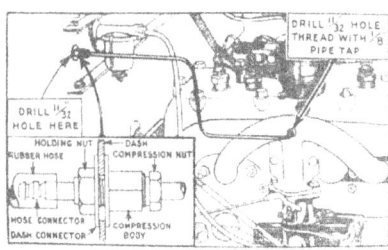

Fig. 752

All Model "A" Tudors are now equipped with a vacuum type windshield wiper.

With the adoption of this type wiper in production on the Tudor we are also supplying it through service so that dealers will be in position to furnish either the electric or vacuum type wipers for replacements.

The vacuum type wiper, complete with necessary fittings for installation, is packed in a carton and listed under part A-17500. List price $5.00, subject to dealer's regular parts discount.

This type of wiper can also be installed on the Fordor and Coupe should an owner prefer it.

HOW IT IS INSTALLED

1. Remove present wiper from the windshield.
2. Insert windshield cleaner control stem through the hole in the frame formerly used for cleaner shaft.
   Place two flat-sided spacers over the stem from the inside of the car.
   Note: On some of the former types of Fordor, only the shorter flat-sided spacer is used.
3. Assemble $\frac{1}{4}$" machine screw from the inside of the car, screwing it into the threaded hole in the cleaner holding plate.
4. Drill the intake manifold with $\frac{11}{32}$" drill $2\frac{1}{2}$" above the carburetor flange and thread with a $\frac{1}{8}$" pipe tap (see Fig. 752).
   Note: On some cars this hole is already provided. In which case simply remove the plug.
5. Screw the compression body into the threaded manifold hole.

Caution: Do not tighten the compression nut which is a part of the body.

6. Drill a $\frac{11}{32}$" hole through the dash to the right of the sediment bulb (see Fig. 752), care must be used not to drill through into the tank.
   Note: This hole is also provided on some cars. In this case, discard the small button used to plug the hole.
7. Insert special connector from the motor side of the dash, holding in position with nut provided.

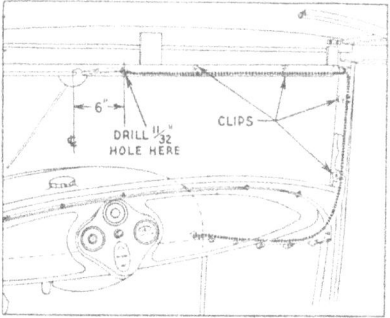

Fig. 753

8. Insert the brass tube end into the compression body as far as possible, and tighten the nut. (Be sure the tube is well seated in the fitting before tightening nut.)
   Connect other end of tube into the dash connector in the same manner.
9. From the inside of the car, slip one end of the hose over connector on dash and carry other end up the right pillar post to the top of the windshield. (Secure hose by clips. The clips can be held by any of the present conveniently located screws.) See Fig. 753.
   Use care not to kink the hose in any way.
10. On the former Fordor models, clamp the hose to the beading on pillar post with special "U" shaped clip provided. On these models, drill an $\frac{11}{32}$" hole through the header not more than 1" from the pillar post. Pass the hose through to the outside of the car where the hose holding clips can be secured under the visor screws.

SERVICING OLD TAIL LAMP AND WIRING ASSEMBLIES

Obsoleting the old design tail lamp and tail lamp wiring assemblies necessitated making several changes in the servicing of those parts.

The present design tail lamps A-13407-A (rustless steel) and A-13407-C (black enamel) are equipped with stop and tail light wires extending from the lamp approximately 22″. One end of the extensions is attached to the tail and stop light sockets in the lamp—the opposite ends are connected to the tail lamp wire by means of a connecting plug A-14487, located 22″ ahead of the tail lamp (see Fig. 858).

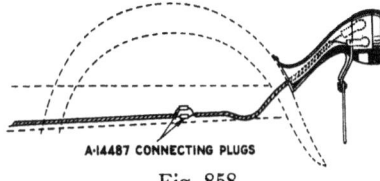

Fig. 858

The adoption of the present design tail lamp with the 22″ wire extensions makes the present tail lamp wire assembly 22″ shorter than the old design.

If after present stocks are exhausted you receive a call for an old style tail lamp wire assembly, supply the new design together with special extension A-14423-BR (see Fig. 859). This extension has been adopted for service so that the present design tail lamp wiring assembly can be installed on jobs equipped with old style tail lamps. The extension is 22″ long and compensates for the difference in length between the old style and new style wiring assemblies.

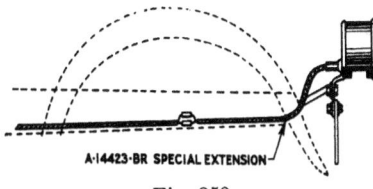

Fig. 859

If after present stocks are exhausted you should receive a call for an old style tail lamp, supply the present design lamp. Due to the 22″ length wires attached to the new lamps, it will be necessary to correspondingly reduce the length of the old style wiring assembly when connecting the extension to the old wiring. This can be easily done by making a loop in the old wiring, taping it together and placing it in the channel of the frame (see Fig. 860).

In replacing an old style tail lamp with the present design lamp, it will of course be necessary to drill the fender to install the present design rear lamp support and reinforcing plate. It will also be necessary to install the present design license bracket.

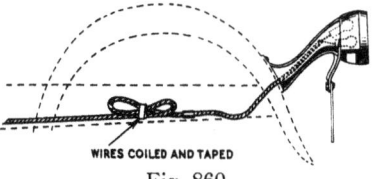

Fig. 860

NEW REAR VIEW MIRROR BRACKET

A few complaints have been received that the rear view mirror was mounted a trifle too high in relation to the back window to secure maximum vision.

To insure 100% vision, the length of the rear view mirror bracket A-17682-B has been changed from 2⅝″ to 3⅜″ (see Fig. 861).

Fig. 861

Ford Service Bulletin *for March*

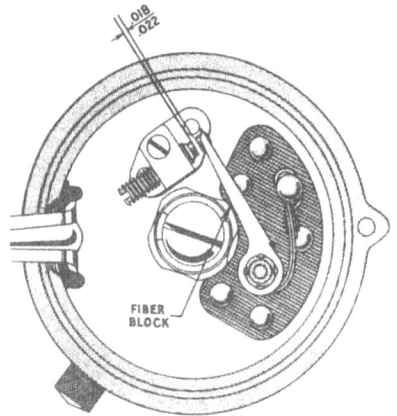

Fig. 856

CHECK BREAKER POINT GAPS

When a new car comes in for inspection, always be sure to check the breaker point gap. This is important. During the first few hundred miles a new car is driven, the contact point on the fiber block on the breaker arm assembly wears slightly until a hard glaze forms on the block. This glaze forms practically a permanent bearing.

During the wearing-in process, the gap between the breaker points becomes slightly less. This is why it is important to check the gap.

Once the fiber block has obtained its permanent bearing, there should be no occasion for further adjustment for some time. A little vaseline, however, should be placed on the distributor cam every 2000 miles.

The gap between the breaker points should measure between .018 and .022 (see Fig. 856).

WINDSHIELD WIPER BLADE

The speed of the vacuum type windshield wiper blade can be regulated by means of the operating switch rod. For example, pushing the switch rod all the way in gives the maximum wiping speed to the wiper blade. Pushing the rod only part way in cuts down the speed of the blade so that it gives a slower wiping action.

RELEASING EMERGENCY BRAKE LEVER

Sometimes owners experience difficulty in releasing the emergency brake lever after the emergency brake has been tightly applied. This is caused by failure to first pull the lever back slightly before pressing down on the release button.

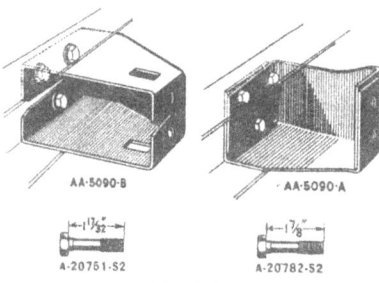

Fig. 857

ENGINE REAR SUPPORT REDESIGNED

The AA-5089-B and 5090-B frame to engine rear supports have been redesigned and are now made from heavier gauge metal.

Redesigning the support eliminated the double thickness of stock formerly used at the inner and outer ends of the old design supports (see Fig. 857). It also changed the length of the frame to engine rear support bolts.

The new design bolt A-20751-S2 is $1\frac{11}{32}$" long. It is not long enough to use with the old design support. The old design bolt A-20758-S2 which was $1\frac{5}{8}$" long has been obsoleted.

If after present stocks are exhausted you should receive any calls for the old style A-20758-S2 bolts, supply bolt A-20782-S2. As this bolt is a trifle longer than the A-20758-S2 bolt, it will be necessary to furnish two A-22217 lock washers with the A-20782-S2 bolt—one of these washers to be slipped over the end of the bolt to compensate for its extra length, the other washer to be used to lock the nut which is used with the bolt, as this bolt is not drilled for a castle nut and cotter key.

SPECIFICATIONS

AXLE, FRONT
Material — Chrome aloy forging.
Caster — 5°
Type — "I" section; reverse Elliott.
Bearings — Taper roller, adjustable.
  Inner Cup — Bower #A-1202. Inner Cone — Timken #A-1201. Outer Cup — Bower #A-1217. Outer Cone — Timken #A-1216.

AXLE, REAR
Material — Ford carbon manganese steel.
Type — Three-quarter floating.
Bearings — Straight roller.
Gear type — Spiral bevel gear.
Gear ratio — 3.70 to 1 (3.77 & 4.11 optional).
Ring gear — 8.4" pitch diameter.
Shaft — 1⅛" diameter.
Pinion bearing — Double taper roller.
Differential bearings — Single taper roller.
Differential gears — Integral with axles.

BATTERY
Type — Ford
Ground — Positive pole ground.
Capacity — 80 ampere-hours; 3 cells; 39 plates.
Charging rate — 10 to 12 amperes.

BRAKES
Service
Type — Ford; mechanical, internal expanding.
Foot pedal — Four-wheels.
Adjustment — Square stud outside operating a wedge.
Percent of braking power —
  Front wheels 40.
  Rear wheels 60.
Brake shoe lining — Woven wire asbestos, 14" long x 1½" wide x 3/16" thick; two shoes per wheel.
Brake drum — 11" diameter x 1¾" wide.
Total braking surface — 168 sq. in.

Parking
Type — Rear brake shoes pick-up.
Hand lever — Located on left frame rail; hand grip release.
Adjustment — Screwed clevis on linkage plus screw wedge.
Total braking surface — 84 sq. in.

CAMSHAFT
Diameter — ⅞"
Bearings —
  Five, each 1-9/16" in diameter
  Length front — 1¾"
  Length second — ⅞"
  Length third — 2"
  Length fourth — ⅞"
  Length fifth — 1"
Cam lift — .302"
Cam gear — Bakelized material; 50 spiral cut teeth.

CARBURETOR
Make — Zenith
Material — Cast iron
Adjustment — Manual by rotating choke rod under dash panel.

CLUTCH
Type — Ford, multiple disc, dry.
Number of discs — 4 driving, 5 driven.

CONNECTING ROD
Material — Steel forging, "X" section
  (Also some welded tubular section).
Length — 7½"
Crank end — Babbitt, 1½" diam. x 1⅝" long
Piston end — Babbitt, 1" diam. x 1⅝" long.

COOLING SYSTEM
Type — Thermo-syphon plus centrifugal pump.
Fan — Two blade "airplane" type, 16" diam.
Drive — ⅝" "V" belt; 1½ times engine speed.
Radiator core — Fin & tube; shrouded for fan.
Radiator shell — Steel, bright nickel finish.
Capacity, system — 3 gallons.
Radiator hose —
  Upper 2" diam. x 6¼" long.
  Lower 1¾" diam. x 2¾" long.

CRANKSHAFT
Material — Ford carbon manganese steel.
Length — 26¼" overall.
Main bearings —
  1⅝" diam. x 2" long, front & center.
  1⅝" diam. x 3" long, rear.
Crank bearings — 1½" diam. x 1⅝" long.
Crank gear — Steel, 25 spiral-cut teeth.

ENGINE
Bore & Stroke — 3⅞" x 4¼".
Horsepower — S.A.E. rating 24.03 H.P.
Horsepower — Brake H.P. 40 at 2200 R.P.M.
Torque — 128 ft. lbs. at 1000 R.P.M.
Displacement — 200.5 cu. in.
Compression Ratio — 4.22 to 1.
Compression — 76 P.S.I., gage.
Firing Order — 1, 2, 4, 3,
Material — Gray iron casting.
Cylinders — 4, cast en bloc; ⅛" offset.
Type — 4 cycle, "L" head, 8 valves on right.
Head — Demountable, held by 12 7/16" studs.
Suspension — 4 point.
Flywheel — Cast iron.
Ring gear — 112 teeth; 14.2" outside diam.
Total Weight — 475 lbs. including clutch & transmission.
Color — Ford Engine Green, Ditzler Co. #DE-40133.

FUEL TANK
Capacity — 11 gallons.

GENERATOR
Type — Ford, "powerhouse"; 6 pole, 5 brush.
Speed — 1½ times engine speed.
Charging Rate — 12 amperes, normal.

IGNITION
Battery — Ford; 6 volts; 80 ampere-hours 39 plates.
Coil — Ford.
Distributor — Ford design eliminating high tension wires to spark plugs.
Breaker Points — Gap .016" to .022".
Spark Plug — ⅞" S.A.E.; Champion No. 3; gap .025 to .030.
Lock — "Electrolock"; theft-proof armored cable to distributor.

LIGHTS
Head Lights
Material — Steel, finished in bright nickel.
Design — Acorn.
Make — Ford.
Lens — Ford; vertical flutes; 8½" diameter.
Bulb — 2 filament; 21 C.P. and 3 C.P.

Tail & Stop Lights
Material — Steel backing plate, brass cover, finished in bright nickel.
Make — Ford "Duolight".

LUBRICATION
Engine
Type — Gear pump to valve chamber; gravity flow to main bearings; splash to other parts.
Oil Capacity — 5 quarts.
Chassis
Type — Alemite pressure grease gun & fittings.

PISTONS
Material — Aluminum.
Length — 3-29/32".
Ring groove width —
　Upper two, 1/8", compression.
　Lower one, 5/32", oil control.
Ring groove depth — 7/32", all.

ROAD CLEARANCE
Road Clearance — 9½" at differential housing.

SHOCK ABSORBERS
Type — Houdaille, double acting, hydraulic.

SPRINGS
Material — Chrome steel.
Type — Transverse.
Front Spring, all cars — 10 leaves; 1¾" wide; 31" free length.
Rear Spring, closed cars — 10 leaves; 2¼" wide; 39" free length.
Rear Spring, open cars — 8 leaves; 2¼" wide; 39½" free length.

STARTING MOTOR
Terminal grounded — Positive.
Normal armature speed — 1500 R.P.M.
Shaft diameter — ½".
Type of drive — Abell.

STEERING GEAR
Type — Worm & 7-tooth sector.
Ratio — 11¼ to 1.
Steering wheel — 17" diameter, red composition.

TIRES
Type — Firestone Balloon, for drop-center rims.
Size — 30 x 4.50 (4.50 x 21).
Pressure — 35 P.S.I., front and rear.

DRIVE
Torque tube.

TRANSMISSION
Type — Selective sliding gear (standard).
Make — Ford.

TREAD
Standard Tread — 56"

TURNING DISTANCE
Radius — 17'
Circle — 34'

VALVES
Arrangement — Vertical, right side.
Material — Carbon chrome nickel.
Lift — .287.
Seat angle — 45°
Spring pressure — 36 lbs.
Stem diameter — 5/16"
Head — Mushroom.
Tappet clearance — .015".

WHEELBASE
Wheelbase — 103½".

Changes Between the 1928 & 1929 Model A Ford

	1928	1929
Engine		
Suspension	4 points	3 points
Camshaft bearings	5	3
Length front	1¾ in.	1¾ in.
Length second	⅞ in.	Omitted
Length third	2 in.	2 in.
Length fourth	⅞ in.	Omitted
Length fifth	1 in.	1 in.
Valve material	Carbon chrome nickel alloy	Chrome Silicon alloy
Spark plug gap	.015 to .020	.025 to .030
Rear Axle		
Gear Ratio	3.7 to 1	3.77 to 1
		4.111 to 1 Optional

FORD SERVICE BULLETIN *for July—August*

Model "A" and "AA" Parts Numbering System

For convenience in the listing and ordering of Model "A" and "AA" parts, groups of numbers are assigned to the various assemblies. Parts used on passenger cars carry the prefix "A". Parts used an trucks carry the prefix "AA". When a part is used on both the passenger car and truck it carries the prefix "A". Each assembly and its part fall within a definite numerical group. For example all wheel parts are grouped into numbers between 1000 and 1099, all brake parts between 2000 to 2999, etc. Each of these groups are illustrated below.

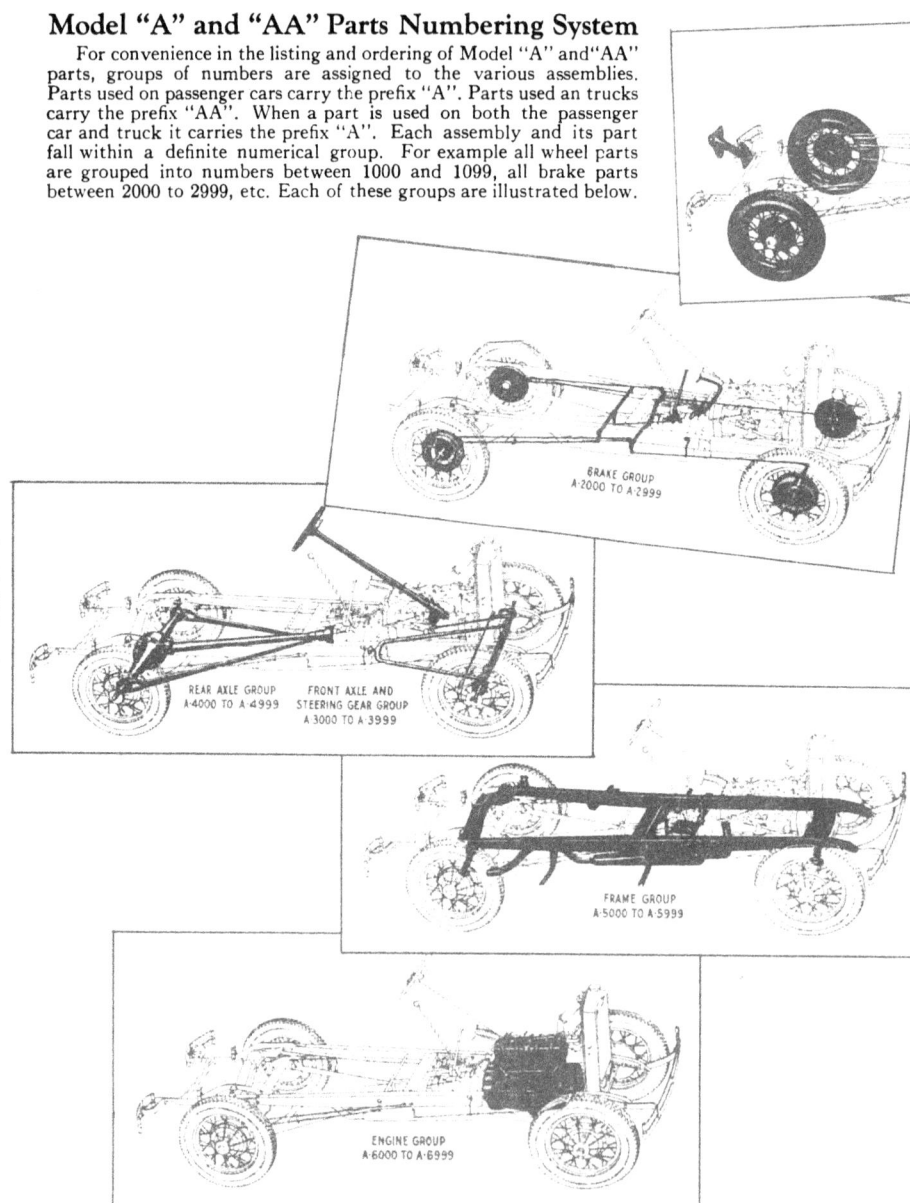

FORD SERVICE BULLETIN *for July—August*

Fig. 1209

Ford Service Bulletin *for July—August*

Springs Used Under Model A and AA Chasses

So that everyone will have a clear understanding of the various springs used under the model A and AA chasses also the body types under which they are installed, we are illustrating below all model A and AA chasses springs.

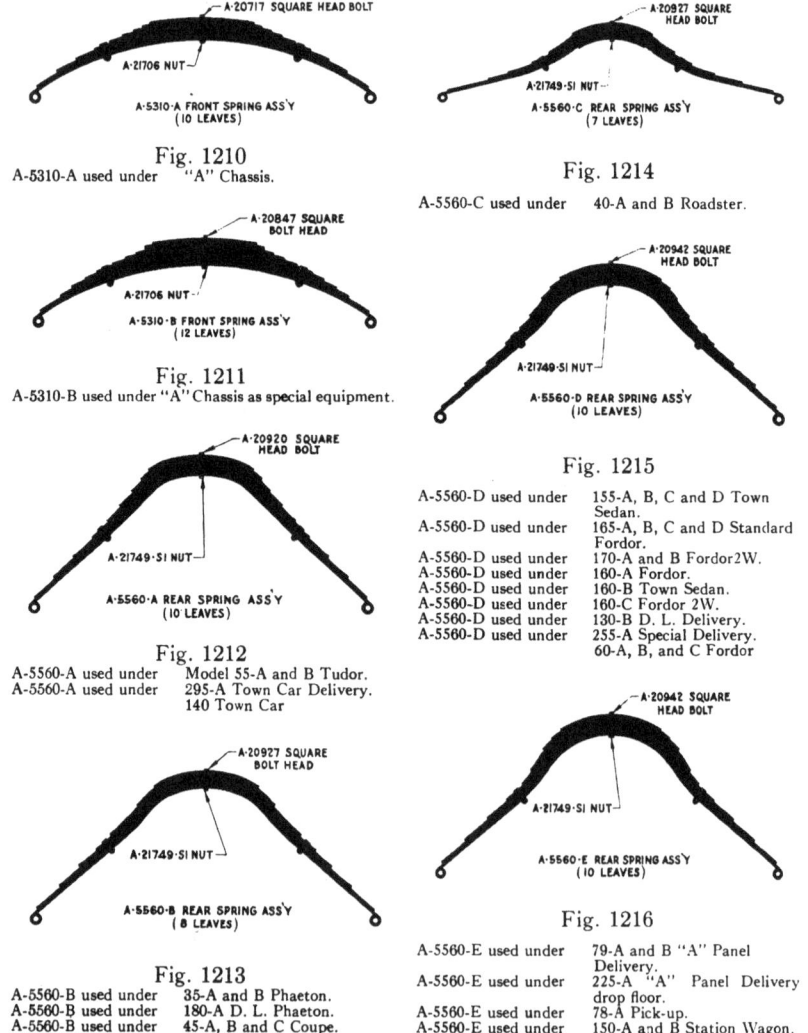

Fig. 1210
A-5310-A used under "A" Chassis.

Fig. 1211
A-5310-B used under "A" Chassis as special equipment.

Fig. 1212
A-5560-A used under Model 55-A and B Tudor.
A-5560-A used under 295-A Town Car Delivery.
140 Town Car

Fig. 1213
A-5560-B used under 35-A and B Phaeton.
A-5560-B used under 180-A D. L. Phaeton.
A-5560-B used under 45-A, B and C Coupe.
A-5560-B used under 50-A and B Sport Coupe.
A-5560-B used under 68-A and B Cabriolet.
A-5560-B used under 190-A Victoria.

Fig. 1214
A-5560-C used under 40-A and B Roadster.

Fig. 1215
A-5560-D used under 155-A, B, C and D Town Sedan.
A-5560-D used under 165-A, B, C and D Standard Fordor.
A-5560-D used under 170-A and B Fordor 2W.
A-5560-D used under 160-A Fordor.
A-5560-D used under 160-B Town Sedan.
A-5560-D used under 160-C Fordor 2W.
A-5560-D used under 130-B D. L. Delivery.
A-5560-D used under 255-A Special Delivery.
60-A, B, and C Fordor

Fig. 1216
A-5560-E used under 79-A and B "A" Panel Delivery.
A-5560-E used under 225-A "A" Panel Delivery drop floor.
A-5560-E used under 78-A Pick-up.
A-5560-E used under 150-A and B Station Wagon.
A-5560-E used under 66-A DeLuxe Pick-up.
76-A and B Open Cab.
82-A and B Closed Cab

Ford Service Bulletin *for July—August*

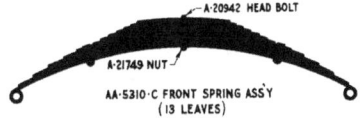

Fig. 1217

AA-5310-C used under "AA" Chassis. (131½" wheelbase.)

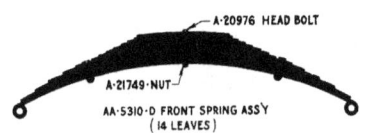

Fig. 1218

AA-5310-D used under 157" wheelbase chassis and as special equipment under the 131½" chassis.

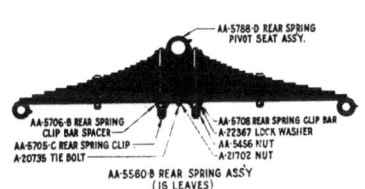

Fig. 1219

NOTE: Truck rear springs as supplied through service include only one spring clip. The spring seat and the second spring clip are not included.

AA-5560-B used under 76-B Open Cab.
AA-5560-B used under 82-B Closed Cab.
AA-5560-B used under 185-B Platform 157" W. B.
AA-5560-B used under 210-A Panel Delivery.
AA-5560-B used under 229-A Service Car.
AA-5560-B used under 187-A Platform 131½" W. B.
AA-5560-B used under 242-A Express H. D. 131½" W. B.
AA-5560-B used under 199-A Ice Wagon. All dump bodies except 205-A.

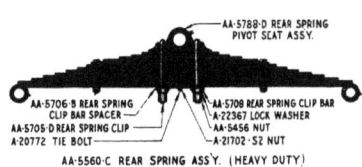

Fig. 1220

AA-5560-C used under 290-A Patrol.
AA-5560-C used under 285-A D. L. Patrol.
AA-5560-C used under 300-A D. L. Delivery.
AA-5560-C used under 195-A Express 131½" W. B.
AA-5560-C used under 85-A and B Panel Delivery 131½" W. B.
AA-5560-C used under 197-A Express 157" W. B.

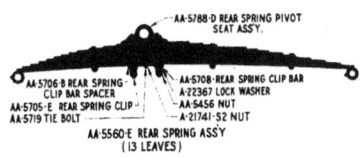

Fig. 1221

AA-5560-E used under 275-A Funeral Coach.
AA-5560-E used under 270-A Funeral Service.
AA-5560-E used under 280-A Ambulance.
AA-5560E special equipment under bodies 85-B, 290-A- 285-A, 300-A.

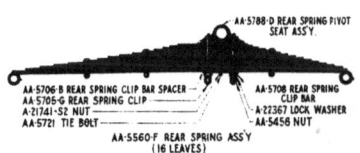

Fig. 1222

AA-5560-F used under 330-A School Bus.

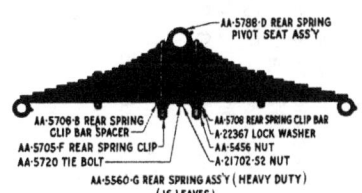

Fig. 1223

AA-5560-G Heavy duty spring used under 205-A Hi-lift dump trucks and as special equipment under "AA" chassis.

97

CLEARANCE LIMITS FOR ASSEMBLY OF "A" ENGINES

(Ford Service Bulletin, June 1931)
(All dimensions in inches)

Piston in cylinders — .002 maximum.
Piston ring gap, lower ring — .008 to .010.
Piston ring gap, center ring — .010 to .012.
Piston ring gap, upper ring — .012 to .015.
Ring groove clearance — .001.
Piston pin fit in connecting rod bushing — .0003 maximum.
Pin in piston — .0002 to .0005 shrink fit.
Pistons assembled with split side toward left side of engine.
Connecting rod side play on crankshaft — .008 to .012.
Connecting rod clearance between piston bosses — .040 to .053.
Connecting rod clearance on crankshaft diameter — .001.
Connecting rods assembled with oil dips toward camshaft.
Crankshaft end play — .002 to .004.
Main bearing clearance — .001.
Camshaft bearing clearance — .003 maximum.
Camshaft end play taken up by tention of spring in front cover. Spring tension approximately 35 lbs.
Valves to push rod clearance — .010 to .013.
Exhaust valves in valve guides — .002.
Intake valves in valve guides — .001 to .0015.
Valve lift — .287.
Push rod clearance — .0015.
Timing gear backlash — .004.
End play of water pump shaft — .006 to .010.
Flywheel eccentricity and wobble (indicator reading) after mounting on crankshaft, not more than .005.
Breaker point gap — .018 to .022.
Spark plug gap — .035.
Free movement or end play in clutch pedal — 1" minimum.

MODEL A FORD ENGINE BOLT SIZES & QUANTITIES

Main Bearings —
   4—½ x 5½ bolts; front & center.
   2—½ x 4 bolts; rear.
Timing Gear Housing —
   6—⅜ x 1⅛ cap screws w/ lockwashers; gear case.
   1—⅜ x 1⅛ cap screw w/ special head for timing ignition (Head 5/16 x ⅝).
Flywheel — 4—7/16 x ¾ SAE cap screws w/ drilled head.
Clutch Housing Plate —
   2—¼ x ½ cap screws w/ lockwashers.
Oil Pan — 20—5/16 x ¾ cap screws w/ lockwashers.
Valve Cover Plate —
   10—5/16 x ¾ cap screws and lockwashers.
Cylinder Head — 12—7/16 SAE nuts.
Oil Return Pipe —
   2—5/16 x 1¼ cap screws w/ copper washers.
Manifold, In. & Exh. —
   4—7/16 SAE nuts w/ thick flat washers.
Starting Motor —
   3—⅜ x 1 cap screws w/ lockwashers; 3 steel shim spacer washers between flange and pad.
Transmission Assembly —
   11—⅜ x 1 cap screws w/ lockwashers.
Steering Column —
   2—½ x 1 SAE cap screws w/ lockwashers; column to chassis.
   2—½ x 1 Fillister head machine screws; cowl support clamp.

COLORS AND COMBINATIONS

The crowning achievement of any restoration must necessarily be the paint job. No matter how meticulous every other detail may have been carried out, an improper or questionable finish will nullify all your good work. On the other hand, nothing causes the exclamations and the expressions of admiration from bystanders like an authentic bit of color.

Of great importance is the striping and, fortunately, it is no longer a lost art. Nearly every wayside village has its hot rod enthusiast who has mastered the tapered brush. A careful perusal of the good-sized photographs reproduced in MODEL A ALBUM will reveal the striping details of every model. The following color schemes for the 1929 model and others will indicate which striping colors were used with each tone grouping. On the following pages are IM-numbers which can be translated into color formulae. Between the two and the Sherwin Williams lacquer mixes, it should be possible to come up with a true duplicate of your car's original finish.

NEW COLOR COMBINATIONS ON BODIES

Phaetons and Roadsters
Bonnie Gray Lower with Chelsea Blue Mouldings and Straw Color Stripe.
Rose Beige Lower with Seal Brown Mouldings and Orange Stripe.
Balsam Green Lower with Valley Green Mouldings and Cream Color Stripe.
Andalusite Blue Lower with Black Mouldings and French Gray Stripe.

Coupes
Bonnie Gray Lower with Chelsea Blue Moulding and Reveals and Straw Stripe.
Vagabond Green Lower with Rockmoss Green Moulding and Reveals and Straw Stripe.
Rose Beige Lower with Seal Brown Moulding and Reveals and Orange Stripe.
Andalusite Blue Lower with Black Moulding and Upper with Niagara Blue Light Reveals and French Gray Stripe.

Tudors
Bonnie Gray Lower with Chelsea Blue Moulding, Reveals and Upper Back with Straw Stripe.
Vagabond Green Lower with Rockmoss Green Moulding, Reveals and Upper Back with Straw Stripe.
Rose Beige Lower with Seal Brown Moulding, Reveals and Upper Back with Orange Stripe.
Andalusite Blue Lower with Black Moulding and Upper Back, Niagara Blue Light Reveals with French Gray Stripe.

Fordors
Bonnie Gray Lower with Chelsea Blue Belt and Upper body with Bonnie Gray Reveals and Straw Stripe.
Vagabond Green Lower with Rockmoss Green Belt and Upper Body with Vagabond Green Reveals and Straw Stripe.
Bramble Brown Lower with Thorne Brown Belt and Upper Body with Bramble Brown Reveals and Neenah Cream Stripe.
Rose Beige Lower with Seal Brown Belt and Upper Body with Rose Beige Reveals and Orange Stripe.
Andalusite Blue Lower with Andalusite Blue Belt and Upper Body with Niagara Blue Light Reveals and French Gray Stripe.

Town Car
Madras Carbuncle Lower with Black Belt and Madras Carbuncle Upper Body with Casino Red Stripe.
Brewster Green Lower with Black Belt and Brewster Green Upper Body with Serpent Green Stripe.
Thorne Brown Lower with Black Belt and Thorne Brown Upper Body and Orange Stripe.
Black Lower with Black Belt and Upper Body with Gold Stripe.

Station Wagon
Natural Wood Body with Manilla Brown Hood, Cowl and Coupe Pillar.

Taxicab
Balsam Green Lower with Black Belt and Medium Cream Reveals and Upper Back with Cream Stripe on Belt and Balsam Green Stripe on Reveals.
Duchess Blue Lower with Black Belt and Medium Cream Reveals and Upper Back with Cream Stripe on Belt and Duchess Blue Stripe on Reveals.

Commercial Bodies
Solid Rockmoss Green with French Gray Stripe and Commercial Gray Spar Varnish to be used on wood parts.

MODEL "A" FORD PAINT LIST as prepared by Ditzler Paint Company:

ALL 1929 MODELS

	Upper	Lower
Tudor and Coupe	Chelsea Blue (IM-120) Rock Moss Green (IM-117) Seal Brown (IM-118) Black	Bonnie Gray (IM-116) Vagabond Green (IM-122) Rose Beige (IM-119) Andalusite Blue (IM-121)
Fordor	Chelsea Blue (IM-120) Rock Moss Green (IM-117) Seal Brown (IM-118) Andalusite Blue (IM-121)	Bonnie Gray (IM-116) Vagabond Green (IM-122) Rose Beige (IM-119) Aandalusite Blue
Cabriolet	Seal Brown (IM-118)	Cigarette Cream (IM-451)
Phaeton and Roadster	Bonnie Gray (IM-116) Rose Beige (IM-119) Andalusite Blue (IM-121) Balsam Green (IM-124)	Bonnie Gray Rose Beige Andalusite Blue Balsam Green
Taxicab	Medium Cream (IM-125) Medium Cream (IM-125)	Duchess Blue (IM-123) Balsam Green (IM-124)
Town Car	Black Black Black	Brewster Green (IM-1017) Mulberry Maroon (IM-1046) Thorne Brown (IM-283)
Town Sedan	Rock Moss Green (IM-117) Rock Moss Green (IM-117)	Vagabond Green (IM-122) Lawn Green (IM-159)
Commercial Jobs	Rock Moss Green (IM-117)	Rock Moss Green (IM-117)

1930 MODELS

	Upper	Lower
Tudor Sedan	Chicle Drab (IM-91) Kewanee Green (IM-546) Black Thorne Brown (IM-283)	Copra Drab (IM-440) Elk Point Green (IM-543) Andalusite Blue (IM-121) Thorne Brown
Two and Three Window Fordor Sedan	Thorne Brown (IM-283) Chicle Drab (IM-91)	Thorne Brown Copra Drab (IM-446)
Sedan	Kewanee Green (IM-546) Black Andalusite Blue (IM-121)	Elk Point Green (IM-543) Ford Maroon (IM-1011) Andalusite Blue
Town Sedan	Black Chicle Drab (IM-91)	Ford Maroon (IM-1011) Copra Drab (IM-440)
Phaeton and Roadster	Thorne Brown (IM-283) Kewanee Green (IM-546) Chicle Drab (IM-91) Andalusite Blue (IM-121)	Thorne Brown Elk Point Green (IM-543) Copra Drab (IM-440) Andalusite Blue
Sport Coupe	Kewanee Green (IM-546) Black Chicle Drab (IM-91) Thorne Brown (IM-283)	Elk Point Green (IM-543) Andalusite Blue (IM-121) Copra Drab (IM-440) Thorne Brown
Standard Coupe	Andalusite Blue (IM-121) Kewanee Green (IM-546) Chicle Drab (IM-91) Thorne Brown	Andalusite Blue Elk Point Green (IM-543) Copra Drab (IM-440) Thorne Brown
Convertible Cabriolet	Andalusite Blue (IM-121) Seal Brown (IM-118) Moleskin Brown Lt. (IM-544) Kewanee Green (IM-546)	Andalusite Blue Bronson Yellow Elk Point Green (IM-543)

1931 MODELS

	Upper	Lower
Tudor and Standard Fordor	Black	Lonbard Blue (IM-1009)
Standard Coupe and Sport Coupe	Thorne Brown (IM-283) Elk Point Green (IM-543) Cobra Drab (IM-440)	Thorne Brown Chicle Drab (IM-91) Chicle Drab
DeLuxe Cabriolet 2-Window Fordor DeLuxe	Elk Point Green (IM-543) Copra Drab (IM-440)	Kewanee Green (IM-546) Chicle Drab (IM-91)
Coupe, DeLuxe Sedan, Town Sedan, Victoria Coupe	Black Black	Brewster Green Medium (IM-1017) Ford Maroon (IM-1011)
DeLuxe Phaeton	Black	Brewster Green Medium (IM-1017)
DeLuxe Roadster	Moulding—Stone Deep Gray (IM-1015) Moulding—Riviera Blue (IM-1013)	Stone Brown (IM-1016) Washington Blue Medium (IM-1014)
Phaeton and Roadster	Black Black Moulding—Elk Point Green (IM-543) Moulding—Copra Drab (IM-440)	Lombard Blue (IM-1009) Thorne Brown (IM-283) Kewanee Green (IM-546) Chicle Drab (IM-91)
Cabriolet	Seal Brown (IM-118) Lombard Blue (IM-1009) Moleskin Brown (IM-544)	Bronson Yellow (IM-545) Lombard Blue Moleskin Brown
Commercial Jobs	Blue Rock Green (IM-1012)	Blue Rock Green

The above (IM Numbers) can be converted by any Ditzler paint dealer into a formula that the desired colors may be made from.

FORD MODEL A COLOR EQUIVALENTS

Color equivalents are based on DuPont finishes available at DuPont refinish distributors throughout the United States. Dulux finishes are air dry enamels while Duco finishes are lacquers. The **original Model A's were finished in pyroxylin lacquers** except for the fenders which were finished in black dipping enamel.
*Available in metallic quality only.

MUNSELL CODE	COLOR	"DUCO"	"DULUX"
5 YR 5/11	Yukon Yellow		93-003
5 YR 6/13	Pegex Orange		93-1021
5 Y 8/5	Medium Cream	725	246-81373
5 Y 7/6	Cream	1559	246-57336
10 BGB 1/3	Blue Rock Green	2204-H*	2204-H*
5 G 2/2	Rock Moss Green	923-G	246-81572-G
5 G 3/4	Balsam Green		93-1855
5 BG 2/4	Vagabond Green		93-81872
5 Y 5/2	Arabian Sand Light		246-31470
10 YGY 3/2	Commercial Drab		93-35648
10 YRY 5/3	Pembroke Gray		None
10 GBG 4/2	Dawn Gray Light		None
5 G 3/1	Bonnie Gray	1293	246-57107
5 R 1/10	Rubelite Red	1497-M	1497-M
5 R 3/14	Vermilion		93-24119
5 PB 00/2	Lombard Blue		246-55106
10 BGB 1/2	Niagara Blue Dark		246-35961
10 BGB 2/4	Niagara Blue Light	783	246-55446
5 BG 2/2	Gun Metal Blue		93-81872
10 BGB 3/4	Duchess Blue	1345-G	246-57118-G
10 YRY 2/2	Mountain Brown		93-3836
10 GBG 2/4	Valley Green		246-34918
10 GYG 2/2	Highland Green		246-34116
10 GYG 3/1	Kewanee Green	785	246-71075
10 GYG 2/1	Elkspointe Green		246-35859
10 GBG 2/3	L'Anse Green Dark		93-6621
5 G 4/3	Lawn Green	662-G	246-62201-G
5 YR 3/5	Phoenix Brown		93-81412
10 YRY 5/4	Manila Brown	837*	202-55505*
5 YR 0/1	Thorne Brown		246-30340
10 YRY 3/1	Copra Drab	884	246-55551
5 Y 3/2	Chickle Drab		None
5 Y 4/2	Arabian Sand Dark		None
10 BGB 2/3	Washington Blue		246-34760
Neutral 1	Stone Gray Deep		246-51252
5 BG 3/2	Dawn Gray Dark	602	246-55134
10 YRY 4/3	Stone Brown		246-35922
5 GY 1/2	Brewster Green		246-54723
10 YRY 0/2	Seal Brown	658	246-60371
5 Y 7/7	Bronson Yellow		None
5 Y 2/1	Moleskin Brown Light		93-6846
5 R 0/3	Ford Maroon		None
10 BPB 00/2	Lombard Blue	914	246-81580
5 GY 3/3	Kewanee Green		93-546
5 YR 4/3	No. 1 Cord and No. 3 Striped Cloth—Light		246-81467
5 YR 3/1	No. 2 Mohair and No. 3 Striped Cloth—Dark		246-50964
5 YR 3/2	No. 4 Mohair		246-81467
10 YRY 4/3	No. 5 Leather, two toned textured: light area		None
10 YRY 0/1	No. 5 Leather, two toned textured: dark area		None
10 YRY 1/3	No. 6 Leather		93-3836

Sherwin-Williams Color Formulae for the following automobile lacquers (stated by volume):

Riviera Blue
41 Parts Bone Black (Opex No. 31111)
33 Parts Prussian Blue (Opex No. 31044)
24 Parts Auto White (Opex No. 31001)
2 Parts Toning Yellow (Opex No. 31163)

* * * *

Washington Blue
61 Parts Prussian Blue (Opex No. 31044)
30 Parts Bone Black (Opex No. 31111)
9 Parts Auto White (Opex No. 31001)
Touch Toning Yellow (Opex No. 31163)

* * * *

Tacoma Cream
88 Parts Auto White (Opex No. 31001)
12 Parts Toning Yellow (Opex No. 31163)
Touch Red Oxide
Touch Ultramarine Blue

Tacoma Cream was used on the wheels, and a double pin stripe in Tacoma Cream ran over the Riviera Blue moulding reveals.

Although these mixes are for lacquer, they can be used for enamel by substituting comparable shades of mixing colors in enamel material.

The 1931 Model A Ford DeLuxe Roadster in the original factory combinations of Riviera Blue, Washington Blue, Tacoma Cream, and Black, was painted as follows:

Washington Blue was used on the hood and body, doors, rumble seat deck lid, etc.

Riviera Blue was used on the moulding reveals along the side of the hood, along tops of doors, and around the body. Two moulding reveals sweep down the rear of the body and along side the rumble seat deck lid.

Black was used on the fenders and splash aprons, bumper brackets, trunk carrier, etc.

COLORS - MODEL A
1928

Phaeton, Tudor Sedan, Roadster, Coupe, Sport Coupe

Niagara Blue (dark or light) body with French Gray belt, reveals, and stripe.

Arabian Sand (dark or light) body with French Gray belt, reveals and stripe.

Dawn Gray (dark or light) body with French Gray belt, reveals, and stripe.

Gun Metal Blue body with French Gray belt, reveals, and stripe.

On open cars the molding carries the same colors that are used on the reveals of closed cars, the stripe also being added.

Tudor Sedan (adopted February, 1928)

Niagara Blue (light) body with Niagara Blue (dark) upper back, belt molding, and reveals, and French Gray Stripe.

Arabian Sand (dark) body with Copra Drab upper back, belt molding, and reveals, and French Gray stripe.

Dawn Gray (dark) body with Gun Metal Blue upper back, belt molding and reveals, and French Gray stripe.

Niagara Blue (dark) body with Niagara Blue (light) upper back, belt molding, and reveals, and French Gray stripe.

Gun Metal Blue body with Black upper back, belt molding, and reveals, and French Gray stripe.

Fordor Sedan (production begun April 27, 1928)

Balsam Green lower body, Pembroke Gray reveals, Valley Green upper molding and belt, and Old Ivory stripe.

Copra Drab lower body, Copra Drab reveals, Seal Brown upper molding and belt, and French Gray Stripe.

Fordor Sedan

Copra Drab discontinued August 3, 1928. Replaced by Rose Beige for body windshield and window reveals, with Seal Brown moldings and Orange stripe. Andalusite Blue for body, windshield, and window reveals, with Arabian Sand (dark) molding and Orange stripe.

1930

Town Sedan

Black, Chicle Drab and Copra Drab, Maroon, Kewanee Green.

Cabriolet

Yellow and Seal Brown, Moleskin, Andalusite Blue, Kewanee Green, Black.

Tudor Sedan, Fordor Sedan (three window), Fordor Sedan Deluxe (two window), Coupe, Deluxe Coupe, Sport Coupe

Andalusite Blue, Kawanee Green, Thorne Brown, Black, Chicle and Copra Drab.

Phaeton and Roadster

Andalusite Blue, Kewanee Green, Thorne Brown, Chicle and Copra Drab, Black.

DeLuxe Phaeton and DeLuxe Roadster

George Washington Blue with Tacoma Cream stripe and wheels.

Raven Black with Aurora Red stripe and wheels.

Stone Brown body and hood, Stone Gray (deep) molding with Tacoma Cream stripe and wheels. (After 8-8-30).

Brewster Green (medium) body and hood, Black molding with Apple Green stripe and wheels. (After 9-25-30).

Lombard Blue body, Hessian Blue stripe and wheels. (After 10-3-30).

DeLuxe Coupe, DeLuxe Sedan, Town Sedan, Cabriolet *(After 9-25-30)*.

Brewster Green (medium) hood, lower body, and window reveals with Black used for belt and above belt, Apple Green stripe and Black wheels.

1931

Town Sedan, DeLuxe Sedan, Victoria Sedan, DeLuxe Coupe, DeLuxe Tudor Sedan

Black upper body with Ford Maroon lower body and English Coach Vermilion stripe.

Black upper body with Brewster Green lower body and Apple Green stripe.

Copra Drab upper body with Chicle Drab lower body and Straw stripe.

Elkpointe Green upper body with Kewanee Green lower body and Apple Green stripe.

Black upper body with Black lower body and Apple Green (deep) stripe.

Standard Fordor Sedan, Tudor Sedan, Coupe

Black upper body with Thorne Brown lower body and Straw stripe.

Black upper body with Lombard Blue lower body and Hessian Blue stripe.

Copra Drab upper body with Chicle Drab lower body and Straw stripe.

Elkpointe Green upper body with Kewanee Green lower body and Apple Green stripe.

Black upper body with Black lower body and Apple Green (deep) stripe.

Convertible Sedan *(released for production May 22, 1931)*

Copra Drab body, hood, and moldings with Chicle Drab reveals, Straw stripe, and Tacoma Cream wheels.

Washington Blue body, hood, and moldings with Riviera Blue reveals and Tacoma Cream stripe and wheels.

Brewster Green (madium) body, hood, and moldings with Tampa Red reveals, Vermilion stripe and Aurora Red wheels.

1931 continued

Phaeton and Roadster

Thorne Brown upper body with Thorne Brown lower body and Straw stripe.

Lombard Blue upper body with Lombard Blue lower body and Hessian Blue stripe.

Copra Drab upper body with Chicle Drab lower body and Straw stripe.

Elkpointe Green upper body with Kewanee Green lower body and Apple Green stripe.

Black upper body with Black lower body and Apple Green (deep) stripe.

DeLuxe Phaeton and DeLuxe Roadster

Riviera Blue upper body with Washington Blue lower body and Tacoma Cream stripe.

Stone Gray (deep) upper body with Stone Brown lower body and Tacoma Cream stripe.

Black upper body with Brewster Green lower body and Apple Green (stripe).

Black upper body with Black lower body and Apple Green (deep) stripe.

Cabriolet

Black upper body with Brewster Green lower body and Apple Green stripe.

Seal Brown upper body with Bronson Yellow lower body and Orange stripe.

Moleskin Brown upper body with Light lower body and French Gray stripe.

Lombard Blue upper body with Lombard Blue lower body and Hessian Blue stripe.

Elkpointe Green upper body with Kewanee Green lower body and Apple Green stripe.

Black upper body with Black lower body and Apple Green (deep) stripe.

Ford Maroon body and deck with Black reveals and English Coach Vermilion stripe.

Note: After June 19, 1931 only three stripe colors were used: Apple Green, Red, or Tacoma Cream.

Interiors — Trim Miscellaneous

ROOF MATERIAL—Black rubberized interlined on all closed cars except Sport Coupe and Cabriolet. Sport Coupe material: two-tone brown-gray, rubber interlined, pyroxylin coated. Cabriolet: tan fabric interlined with rubber. Phaeton: Black rubberized top material.

HARDWARE—External handles: Chromium plated or stainless steel, Interior handles: Nickle plated.

UPHOLSTERY—

Phaeton: Two tone cross cobra grain artificial leather.

Roadster: Two tone cross cobra grain artificial leather, rumble seat, same.

Coupe: Brown check wool cloth. Pockets in doors.

Deluxe Coupe: Mohair; green or taupe. Bedford cord; deep tan.

Sport Coupe: Above belt line; tan cloth. Below belt line; brown check. Rumble seat; two-tone cross cobra grain artificial leather.

Tudor Sedan: Brown check cloth.

Cabriolet: Tan Bedford cord upholstery and trim below belt line. Rumble seat; two-tone cross cobra grain artificial leather.

Deluxe Sedan: Tan Bedford cord; taupe or green mohair.

Town Sedan: Brown or green mohair.

Station Wagon: Two-tone cross cobra grain artificial leather.

TRIM—Mahogany garnish moldings are found in the **DeLuxe Coupe, Deluxe Sedan, Town Sedan.** Silk curtains on the rear quarter windows and a flexible robe rail are found on the **Town Sedan.** The robe rail is also a feature of the **Deluxe Sedan.**

BODY DIMENSIONS AND CUTAWAYS
Model A Chassis Showing Important Dimensions for Mounting Bodies

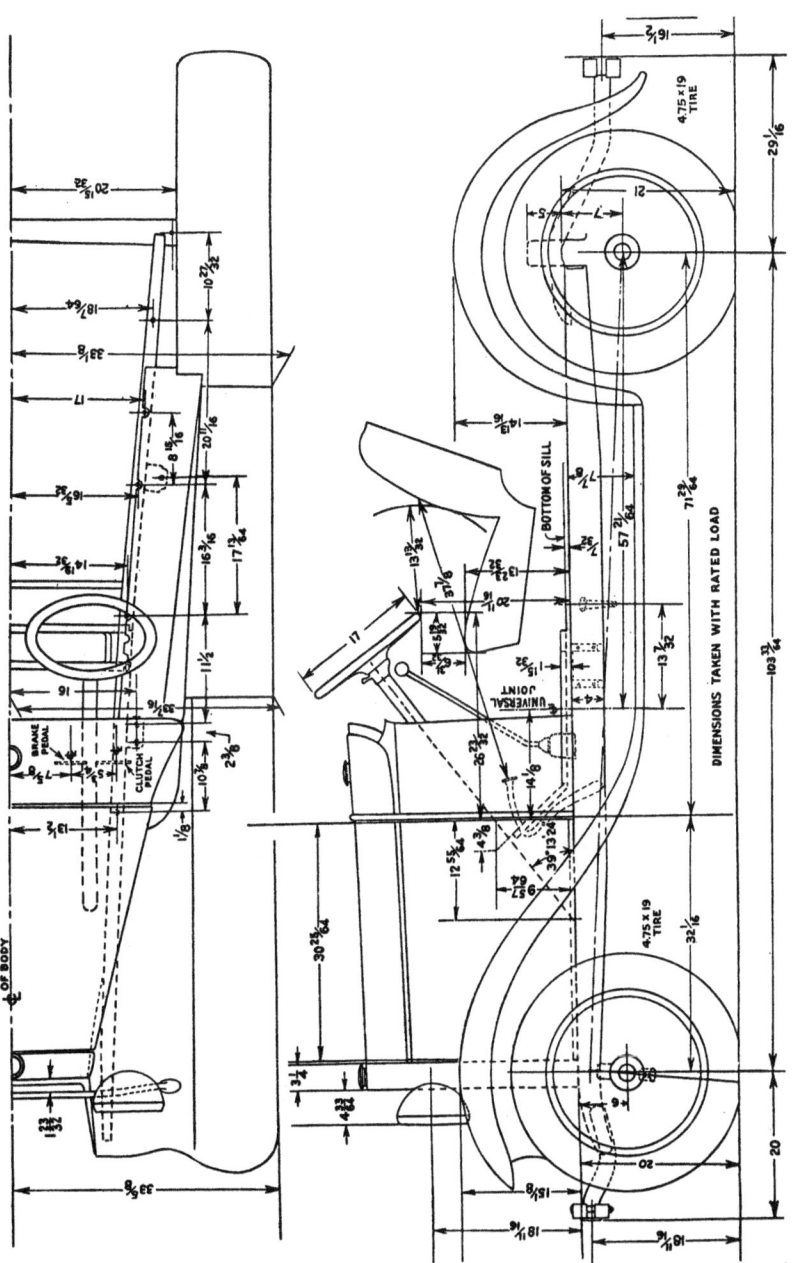

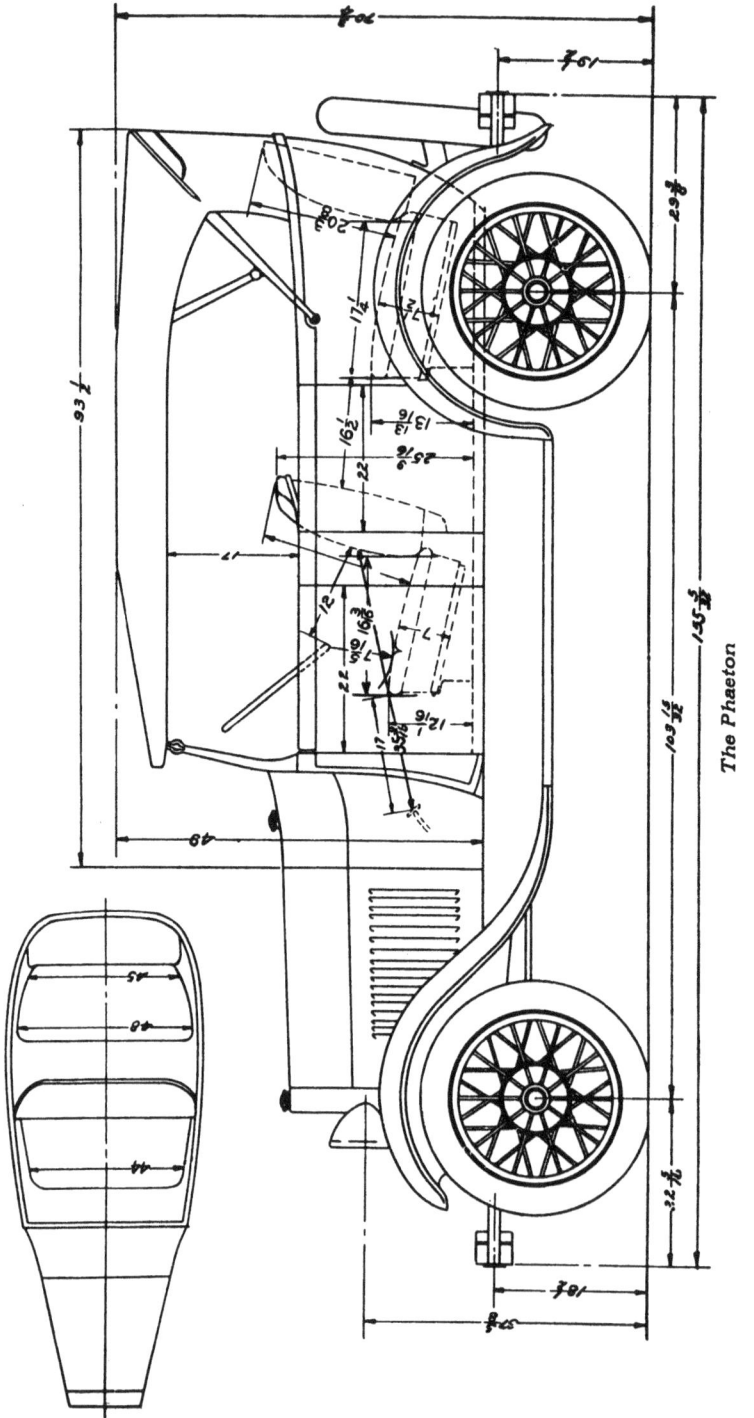

The Phaeton 1929

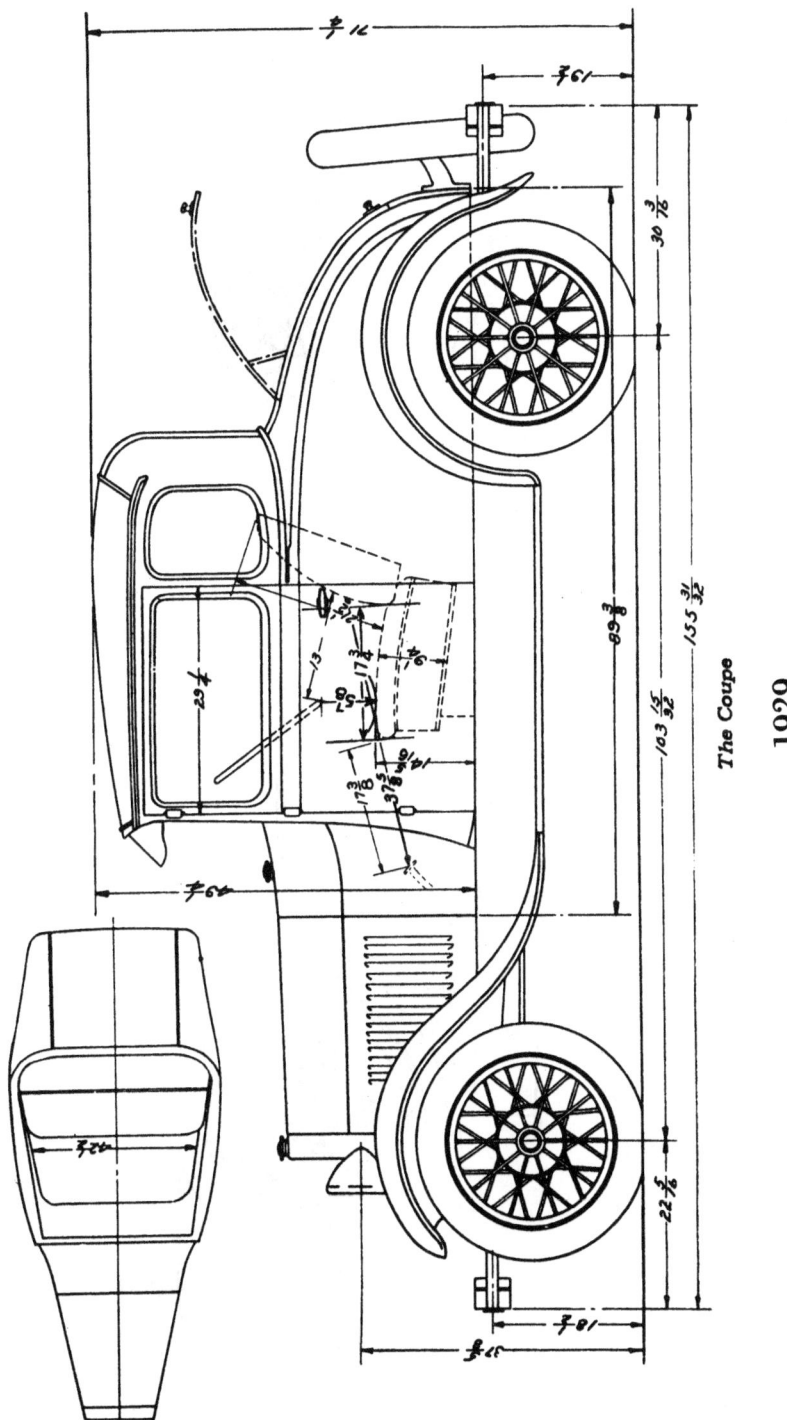

The Sport Coupe 1929

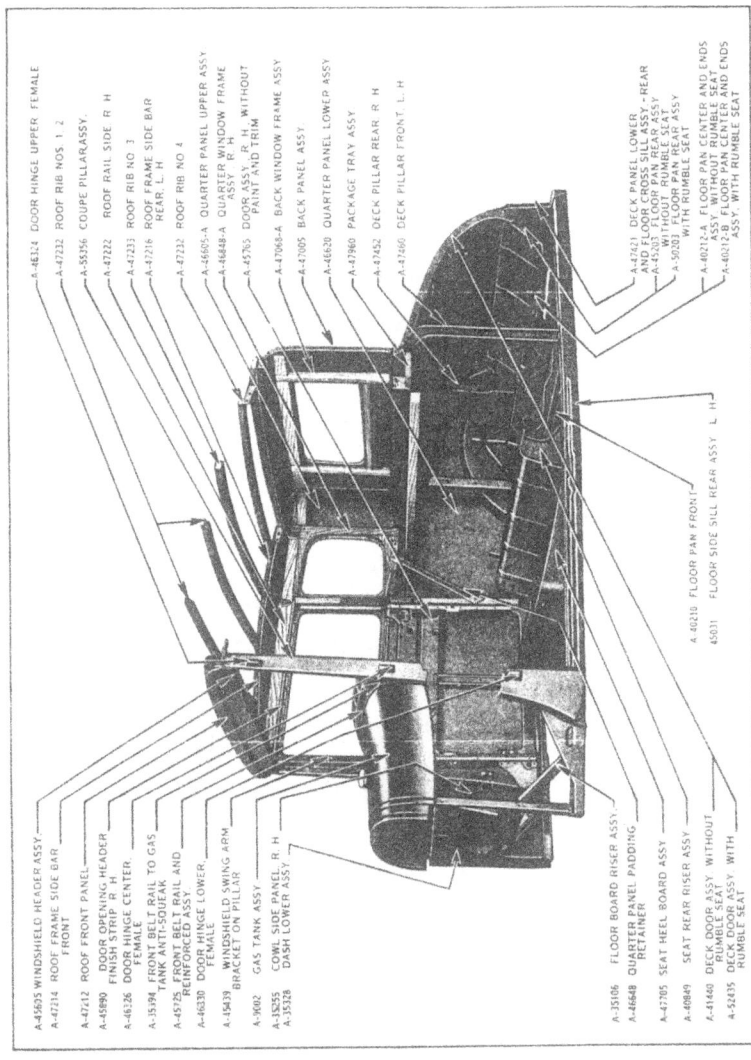

Fig. 783—Sectional View of the Standard Coupe

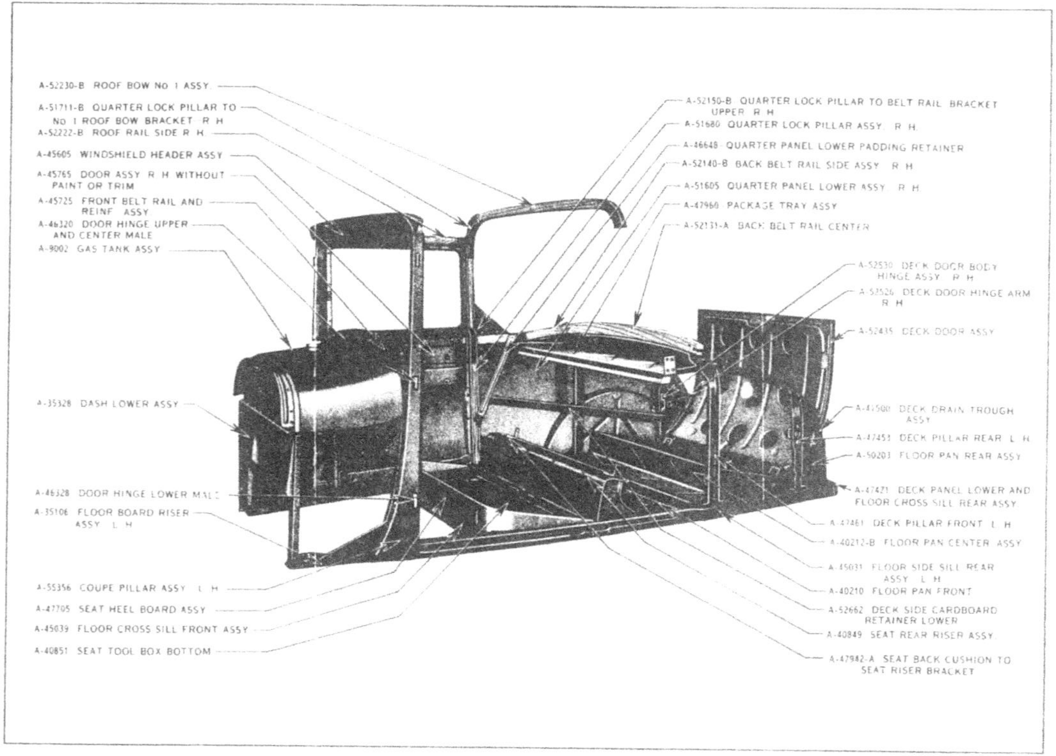

Fig. 800—Sectional View of Sport Coupe

- A-52230-B ROOF BOW NO. 1 ASSY
- A-51711-B QUARTER LOCK PILLAR TO NO. 1 ROOF BOW BRACKET R H
- A-52222-B ROOF RAIL SIDE R H
- A-45605 WINDSHIELD HEADER ASSY
- A-45765 DOOR ASSY R H WITHOUT PAINT OR TRIM
- A-45725 FRONT BELT RAIL AND REINF ASSY
- A-46120 DOOR HINGE UPPER AND CENTER MALE
- A-9002 GAS TANK ASSY

- A-52150-B QUARTER LOCK PILLAR TO BELT RAIL BRACKET UPPER R H
- A-51980 QUARTER LOCK PILLAR ASSY R H
- A-46648 QUARTER PANEL LOWER PADDING RETAINER
- A-52140-B BACK BELT RAIL SIDE ASSY R H
- A-51605 QUARTER PANEL LOWER ASSY R H
- A-47960 PACKAGE TRAY ASSY
- A-52131-A BACK BELT RAIL CENTER

- A-55510 DECK DOOR BODY HINGE ASSY R H
- A-55261 DECK DOOR HINGE ARM R H
- A-55245 DECK DOOR ASSY
- A-43500 DECK DRAIN TROUGH ASSY
- A-47451 DECK PILLAR REAR L H
- A-56001 FLOOR PAN REAR ASSY
- A-47421 DECK PANEL LOWER AND FLOOR CROSS SILL REAR ASSY
- A-47361 DECK PILLAR FRONT L H
- A-46001 FLOOR SIDE SILL REAR ASSY L H
- A-40210 FLOOR PAN FRONT
- A-52662 DECK SIDE CARDBOARD RETAINER LOWER
- A-40849 SEAT REAR RISER ASSY
- A-47982-A SEAT BACK CUSHION TO SEAT RISER BRACKET

- A-35128 DASH LOWER ASSY
- A-46128 DOOR HINGE LOWER MALE
- A-55108 FLOOR BOARD RISER ASSY L H
- A-55356 COUPE PILLAR ASSY L H
- A-47705 SEAT HEEL BOARD ASSY
- A-45639 FLOOR CROSS SILL FRONT ASSY
- A-40851 SEAT TOOL BOX BOTTOM

112

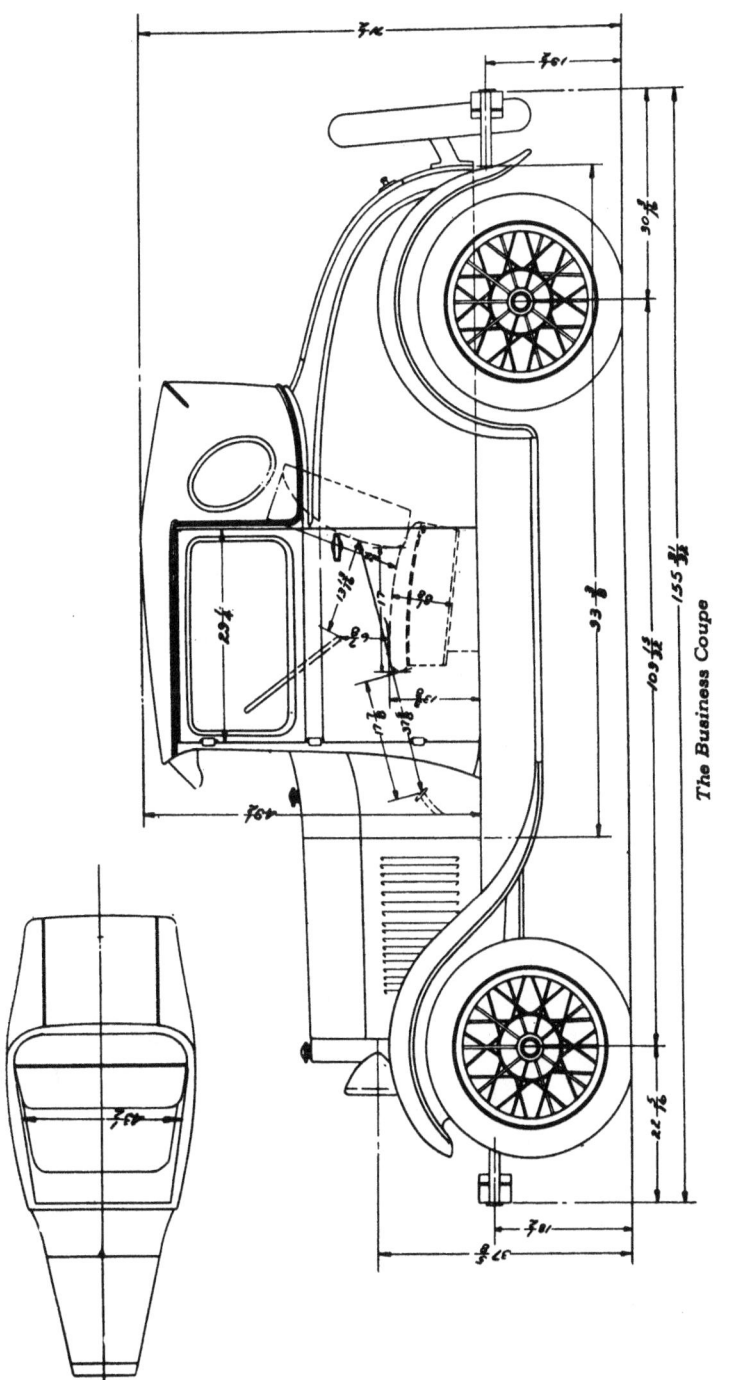

The Business Coupe 1929

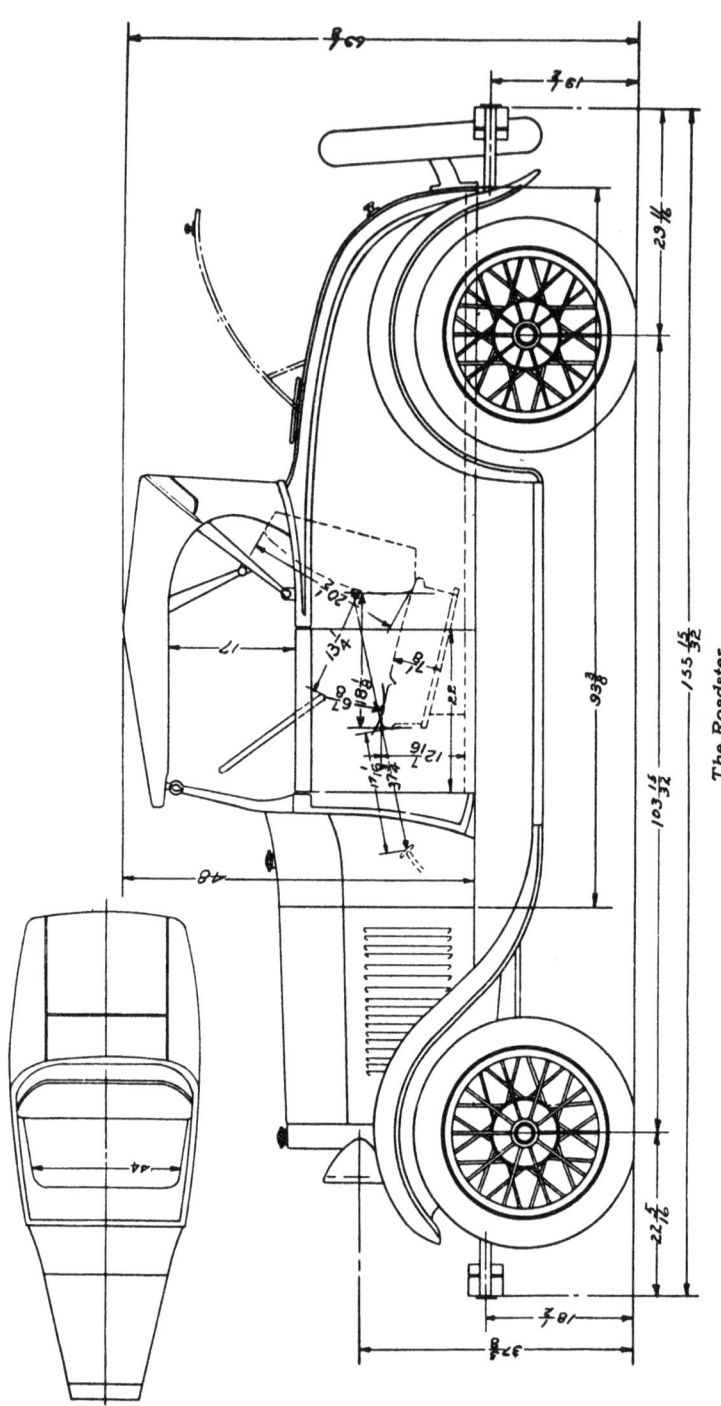

The Roadster 1929

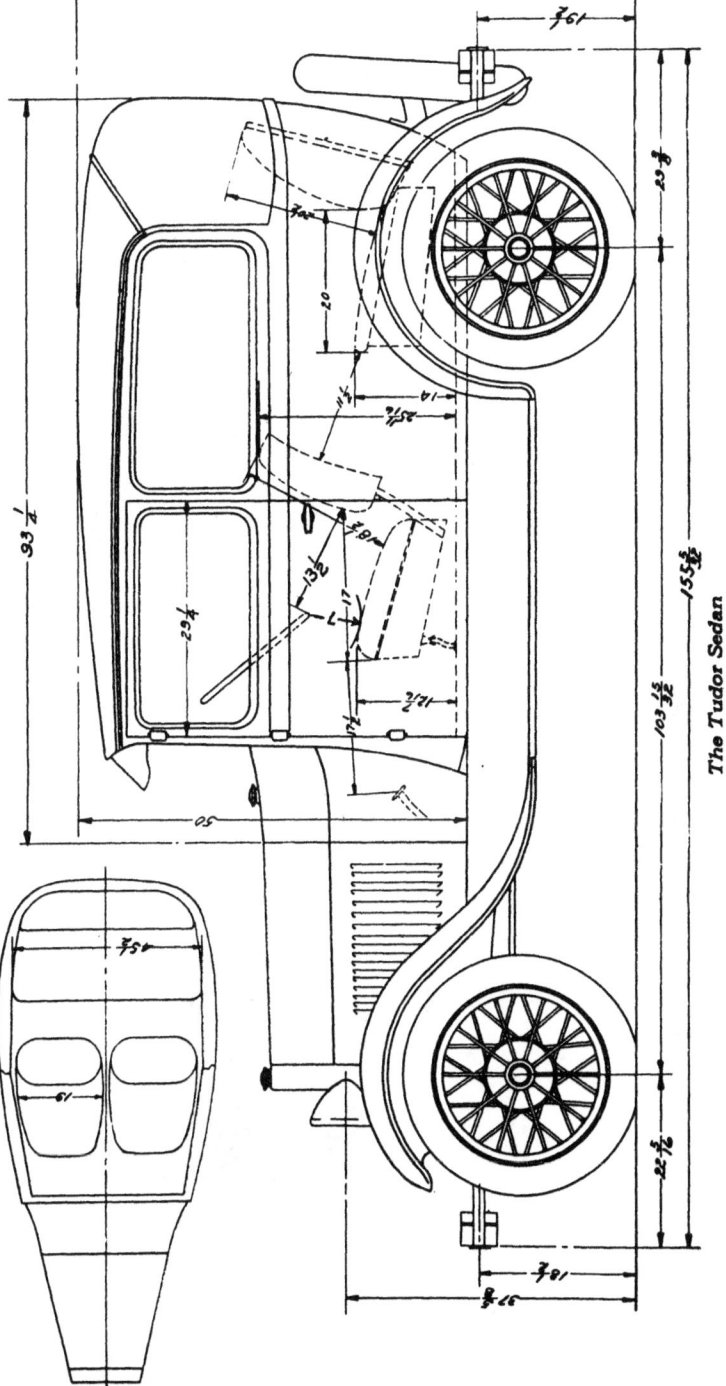

The Tudor Sedan 1929

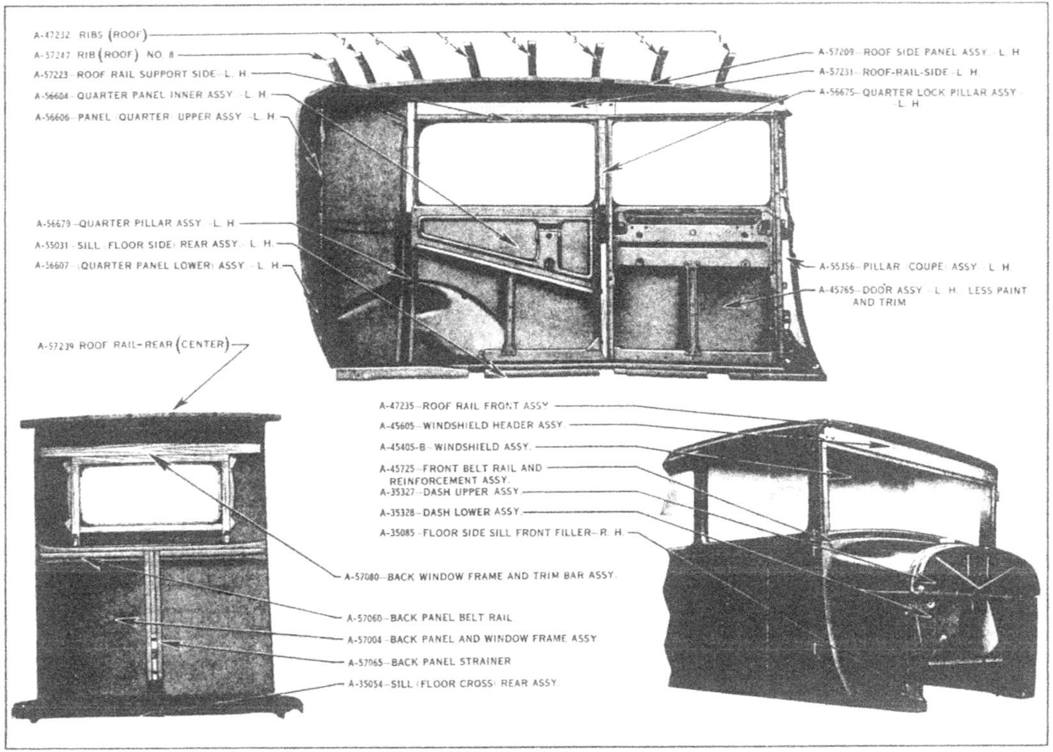

A-47212 - RIBS (ROOF)
A-57247 - RIBS (ROOF) NO. 8
A-57223 - ROOF RAIL SUPPORT SIDE L. H.
A-56604 - QUARTER PANEL INNER ASSY L. H.
A-56606 - PANEL QUARTER UPPER ASSY L. H.

A-56673 - QUARTER PILLAR ASSY L. H.
A-55031 - SILL FLOOR SIDE REAR ASSY L. H.
A-56607 - QUARTER PANEL LOWER ASSY L. H.

A-57239 - ROOF RAIL - REAR (CENTER)

A-57209 - ROOF SIDE PANEL ASSY L. H.
A-57231 - ROOF-RAIL-SIDE L. H.
A-56675 - QUARTER LOCK PILLAR ASSY L. H.

A-55356 - PILLAR COUPE ASSY L. H.
A-45765 - DOOR ASSY L. H. LESS PAINT AND TRIM

A-47235 - ROOF RAIL FRONT ASSY
A-45605 - WINDSHIELD HEADER ASSY
A-45405-B - WINDSHIELD ASSY
A-45725 - FRONT BELT RAIL AND REINFORCEMENT ASSY
A-35327 - DASH UPPER ASSY
A-35128 - DASH LOWER ASSY
A-35085 - FLOOR SIDE SILL FRONT FILLER - R. H.

A-57080 - BACK WINDOW FRAME AND TRIM BAR ASSY
A-57060 - BACK PANEL BELT RAIL
A-57004 - BACK PANEL AND WINDOW FRAME ASSY
A-57065 - BACK PANEL STRAINER
A-35054 - SILL FLOOR CROSS REAR ASSY

Fig. 772—Sectional View of the Tudor

The Fordor Sedan
1929

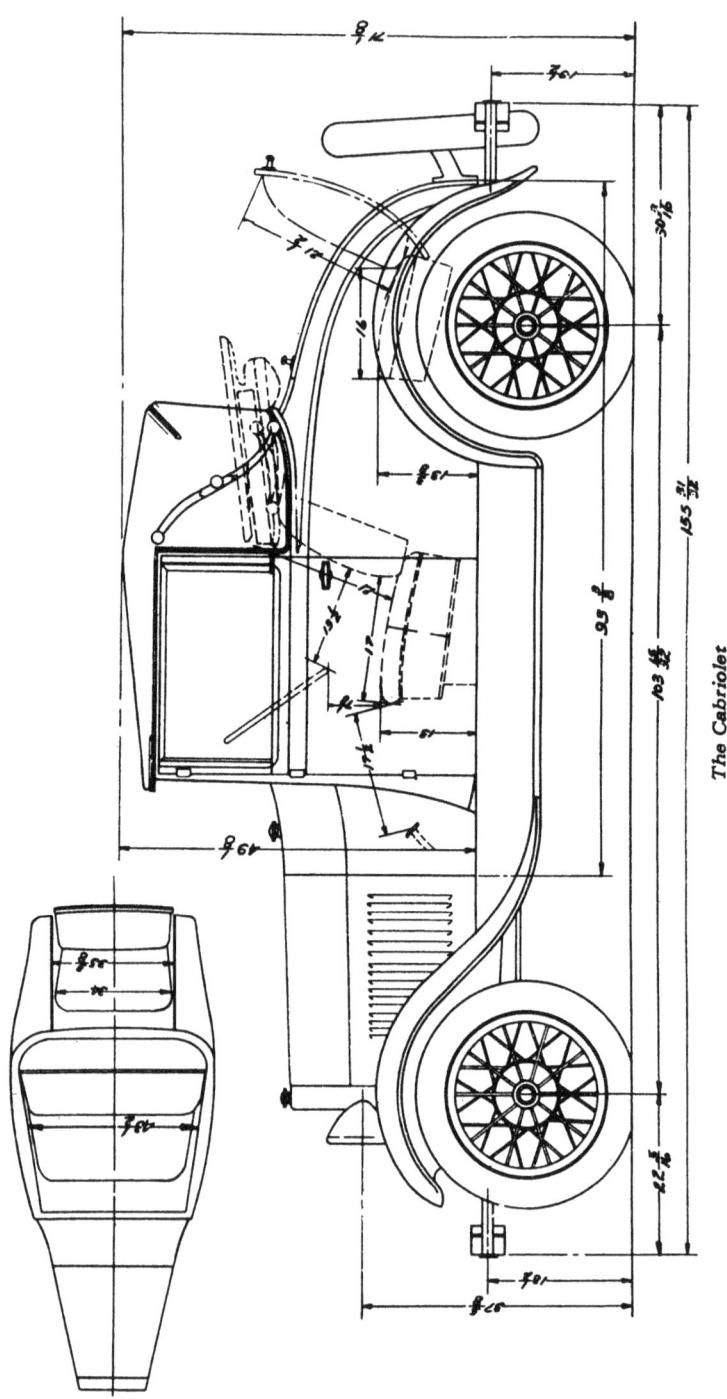

The Cabriolet 1929

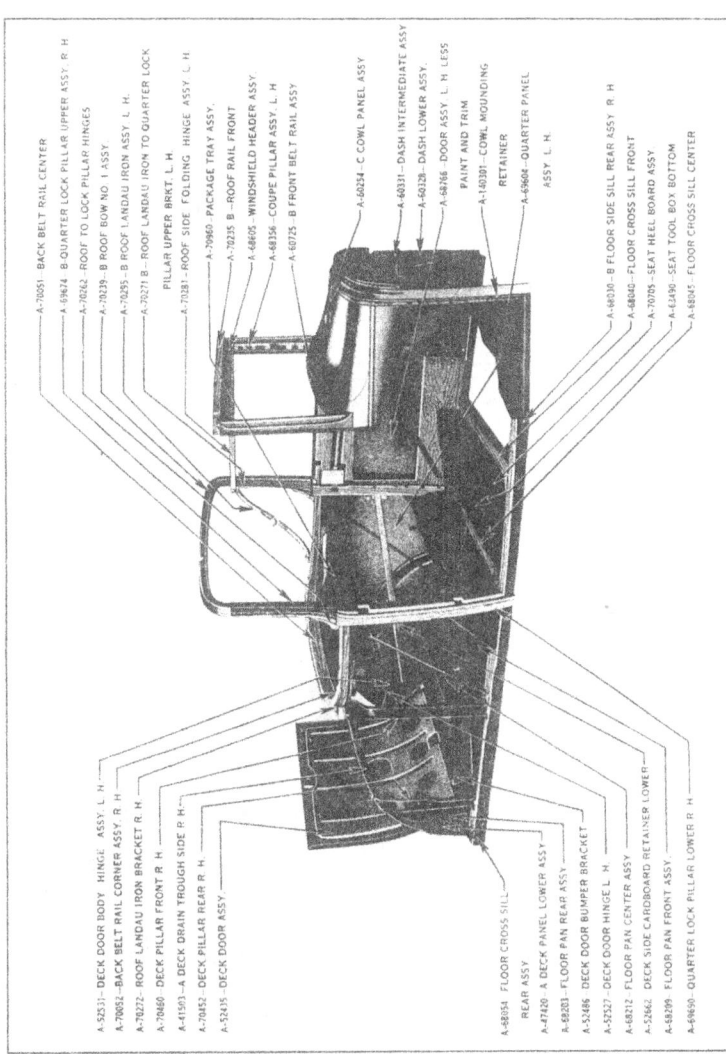

Fig. 812—Sectional View of Cabriolet (1929 design.)

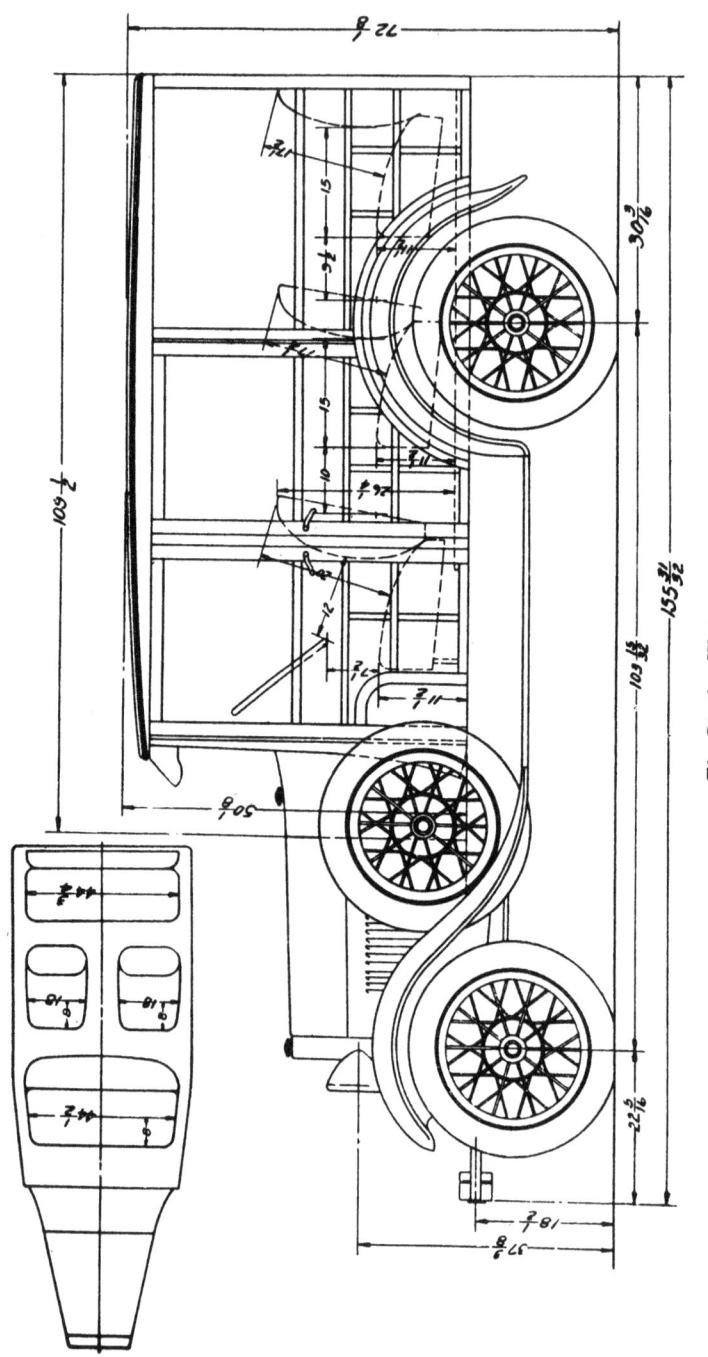

The Station Wagon 1929

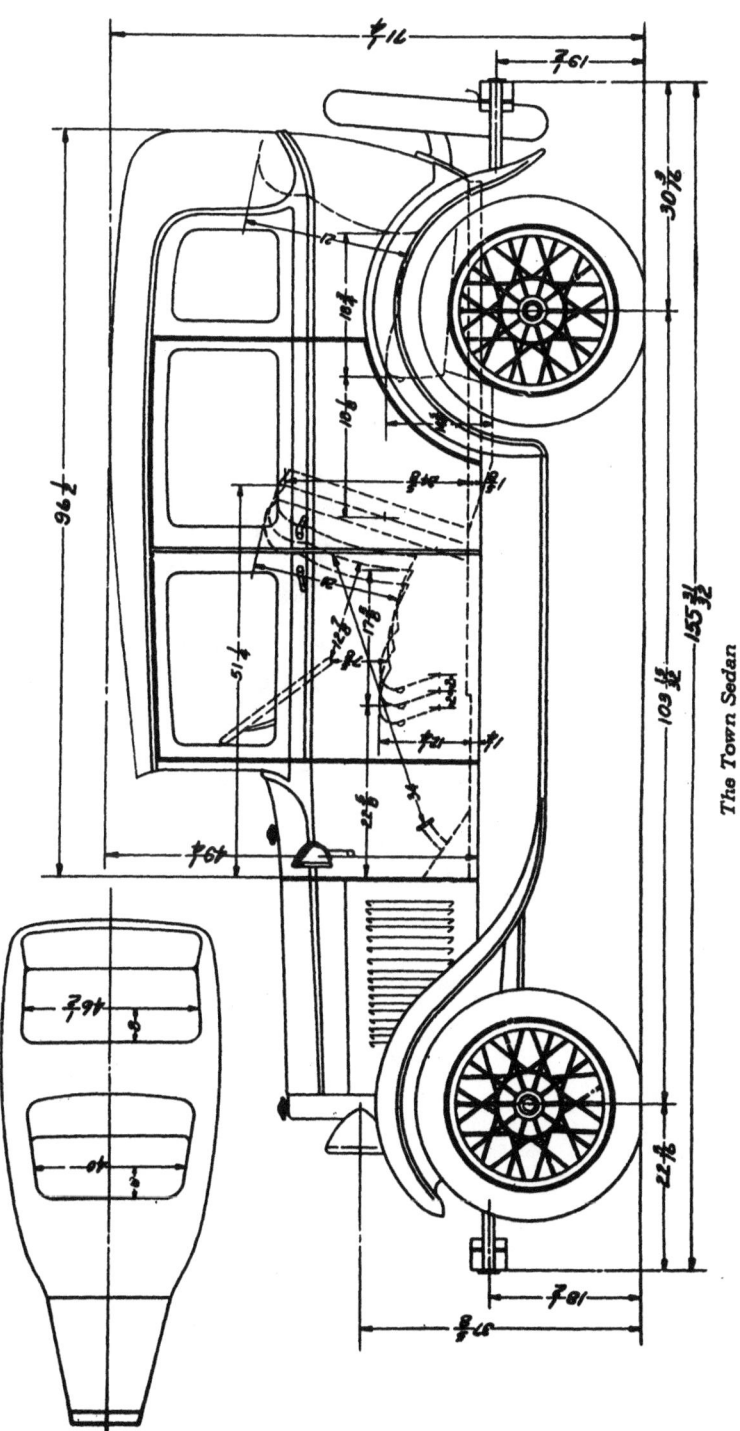

The Town Sedan 1929

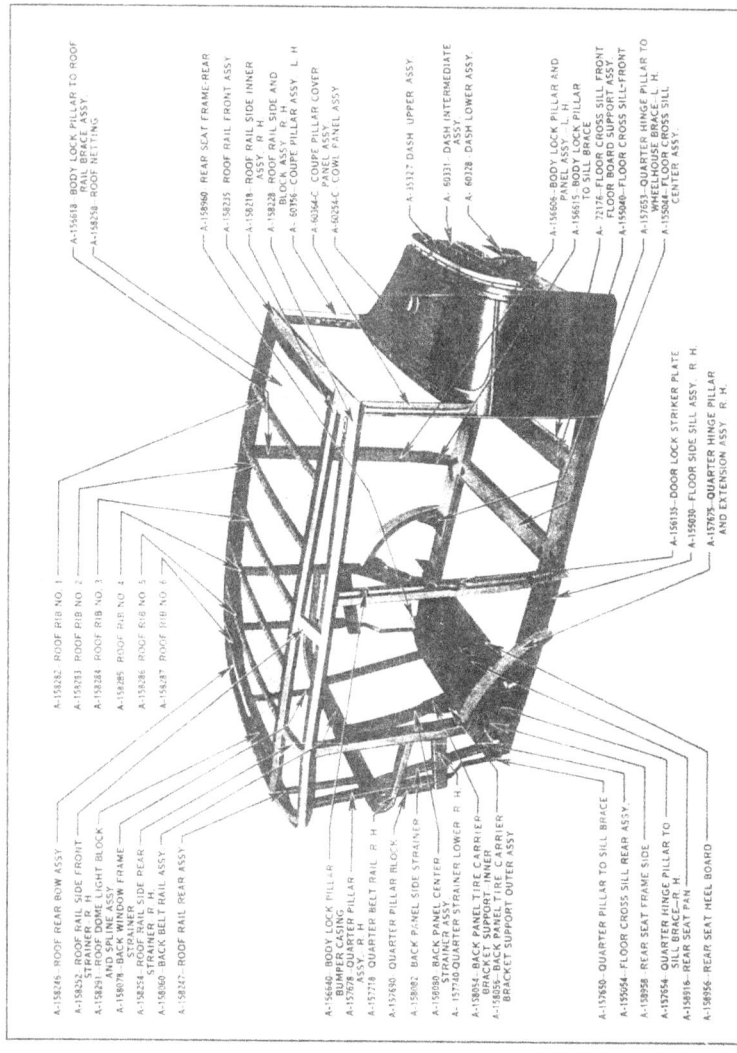

Fig. 756—Skeleton View of Town Sedan

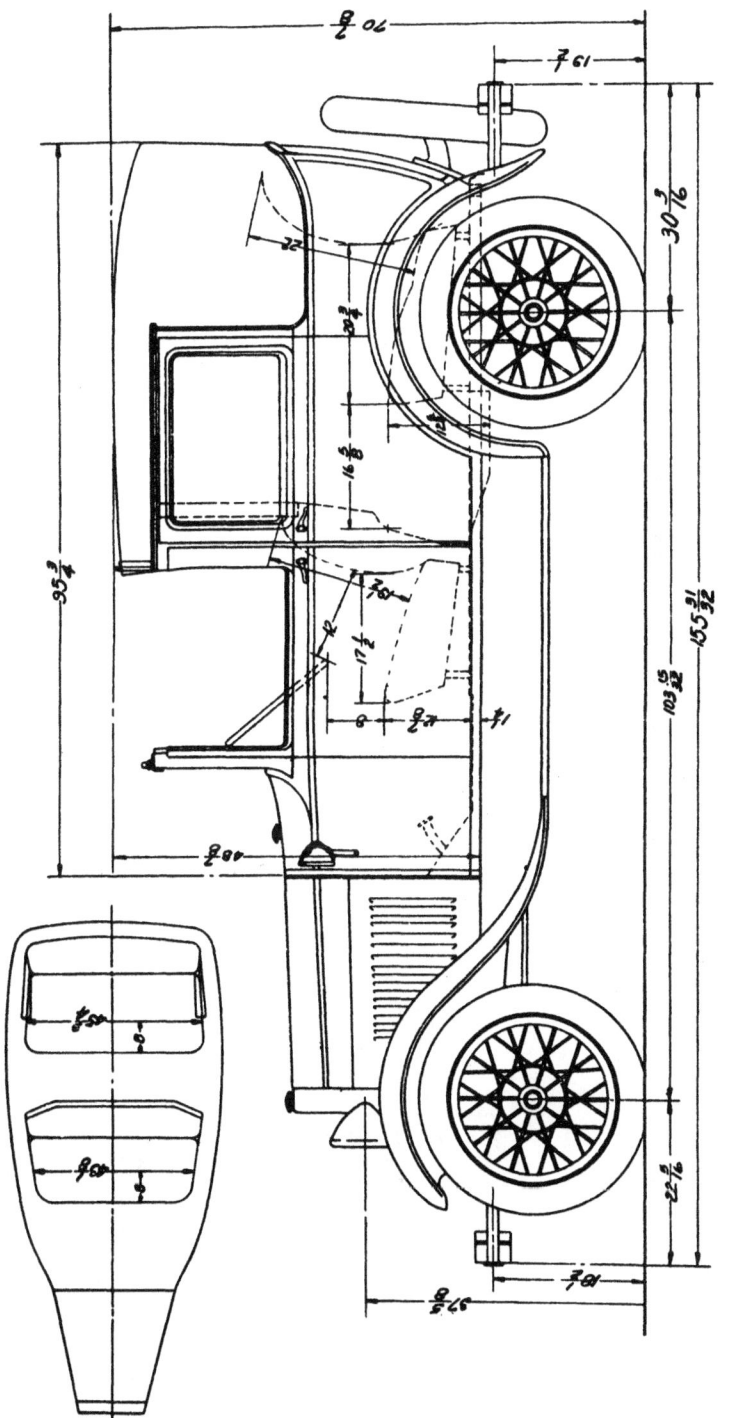

The Town Car

1929

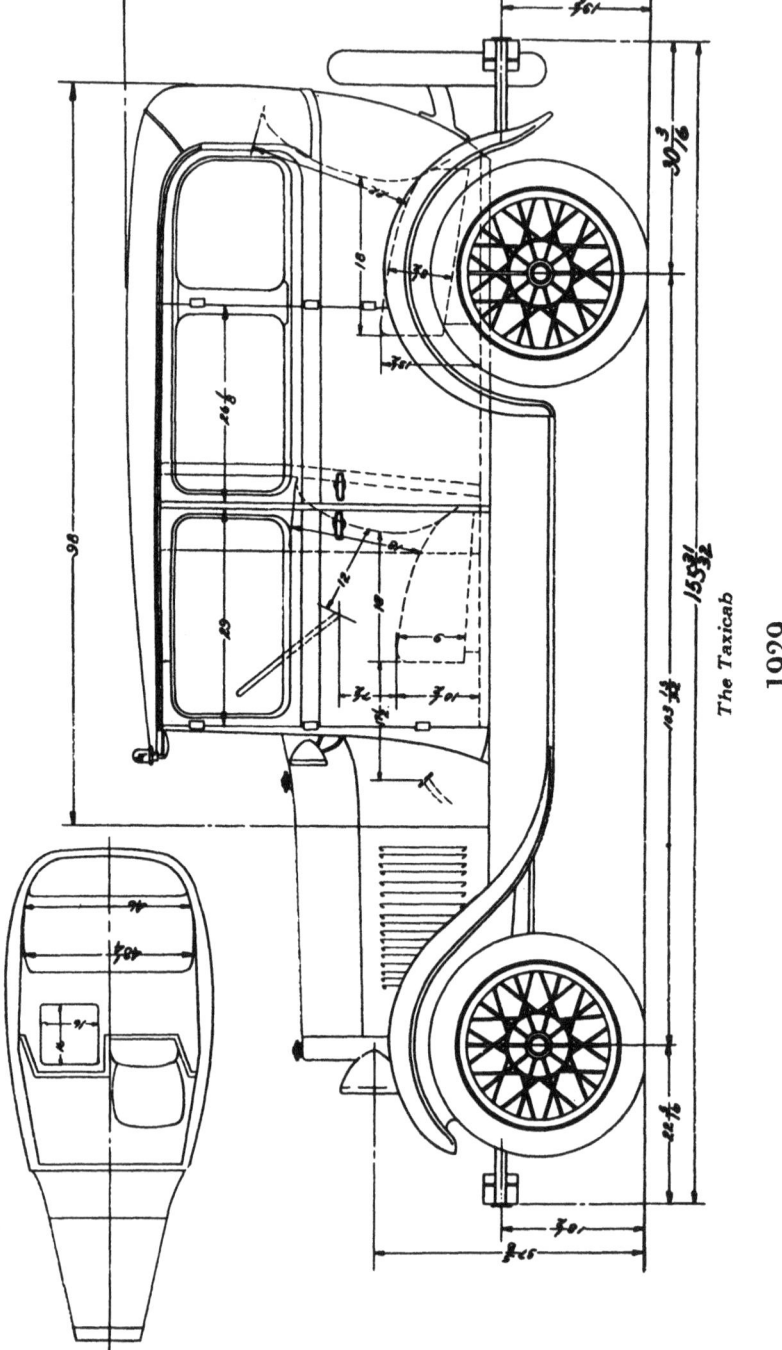

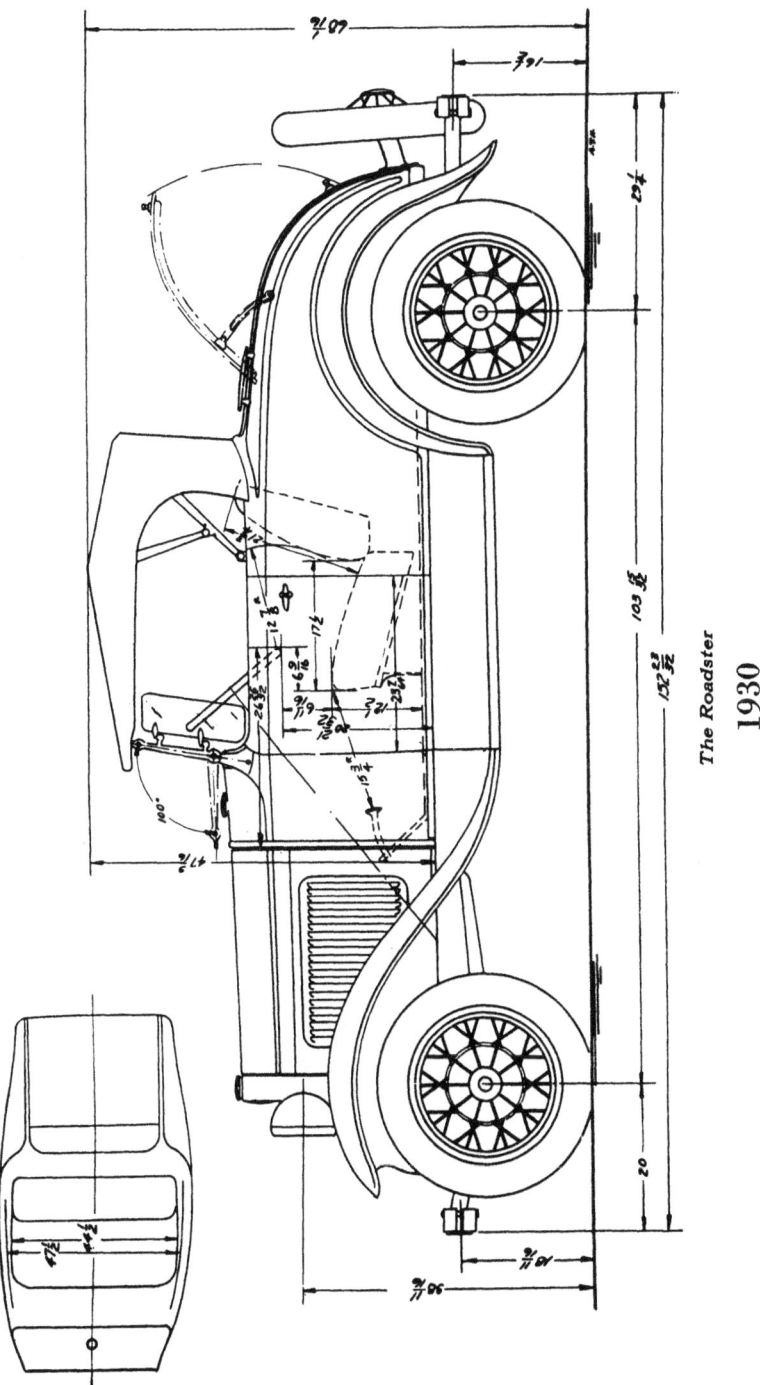

The Roadster
1930

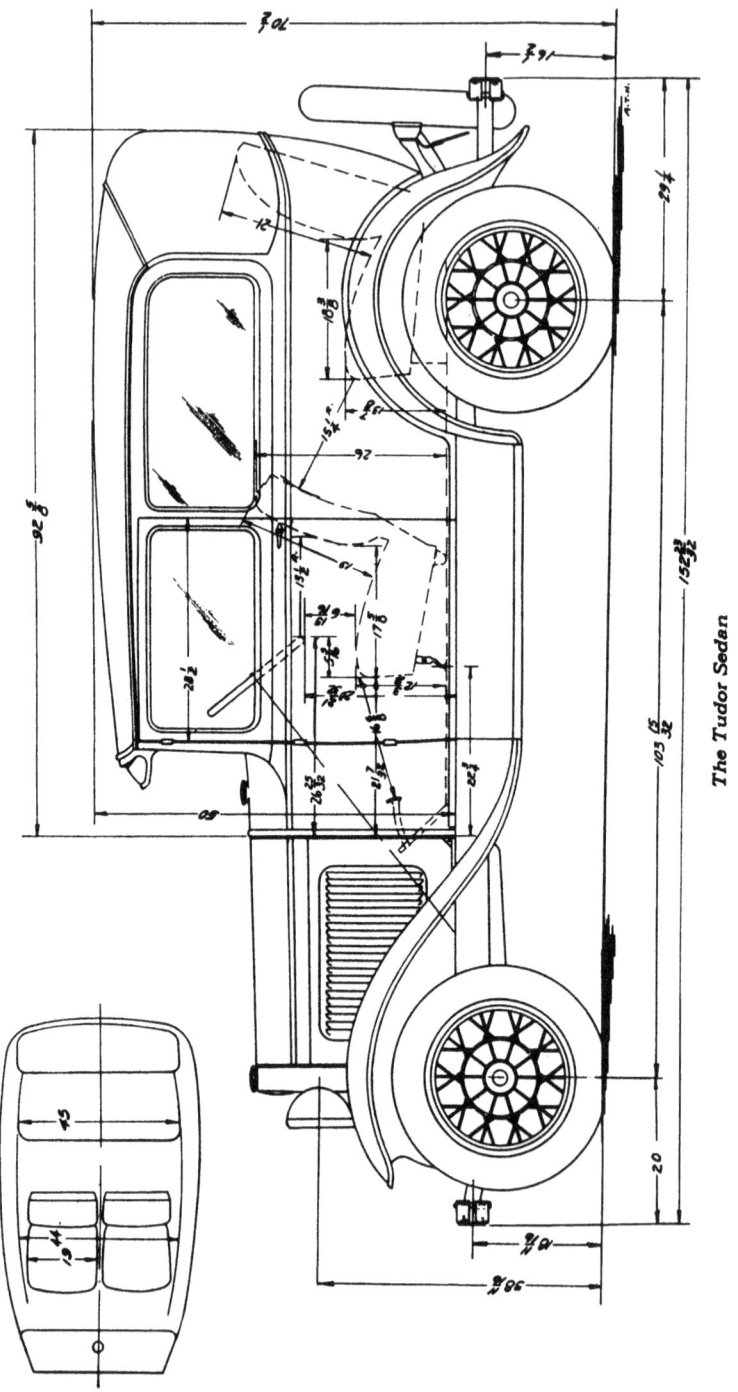

The Tudor Sedan 1930

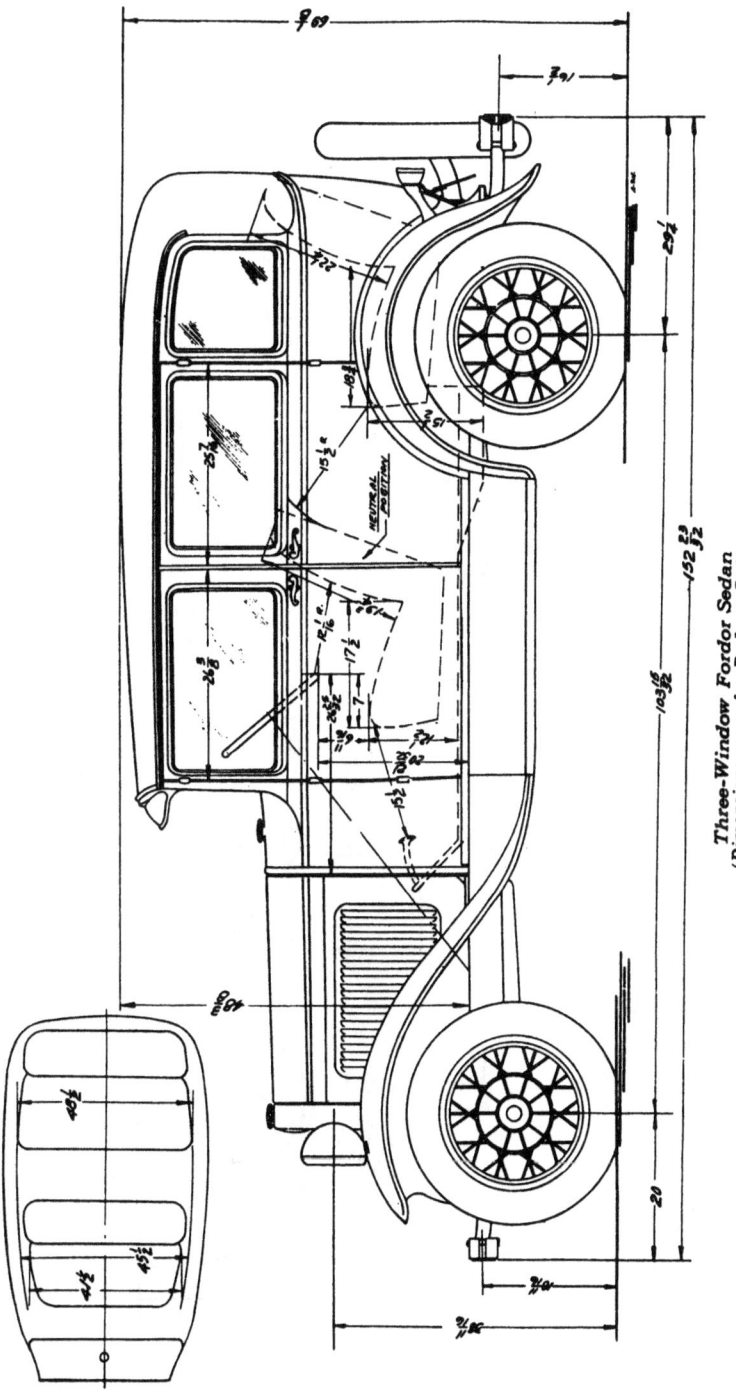

Three-Window Fordor Sedan
(Dimensions same for De Luxe Sedan)
1930

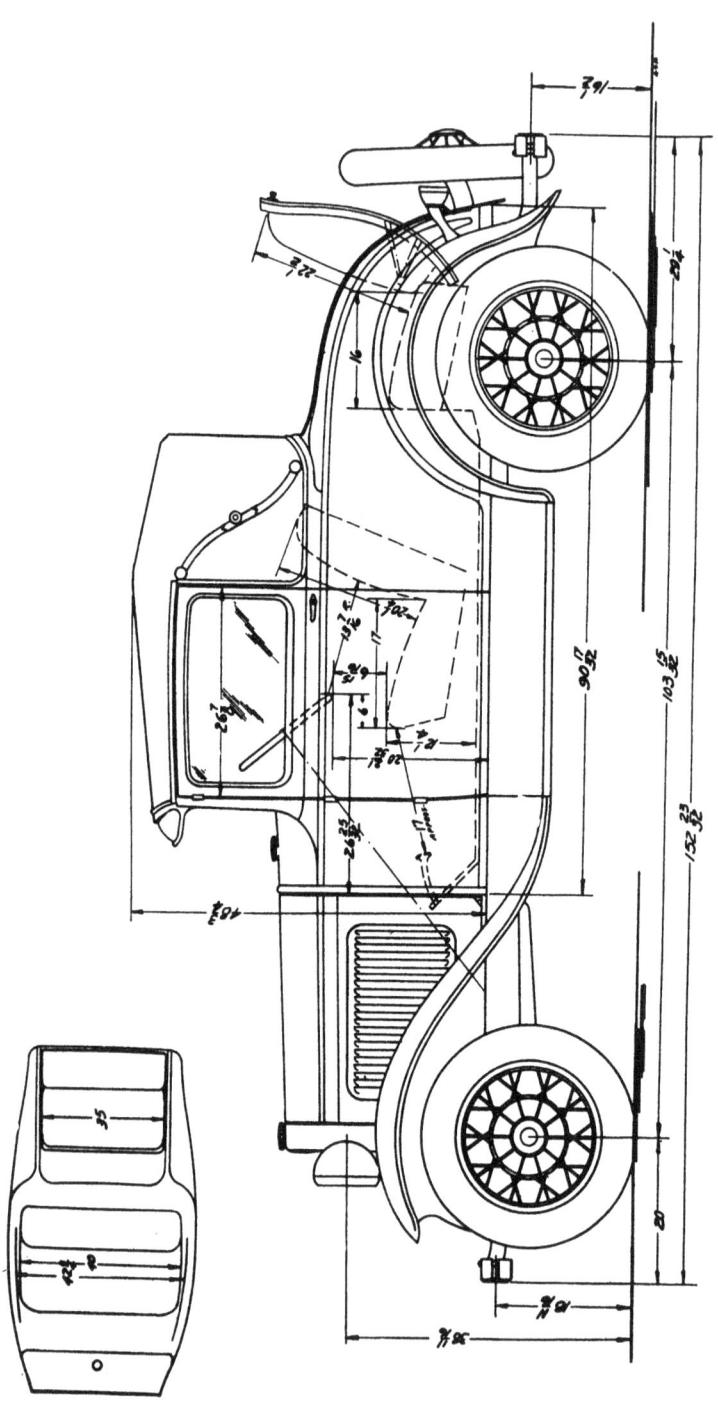

The Sport Coupe
(Dimensions approximately same for Cabriolet)
1930

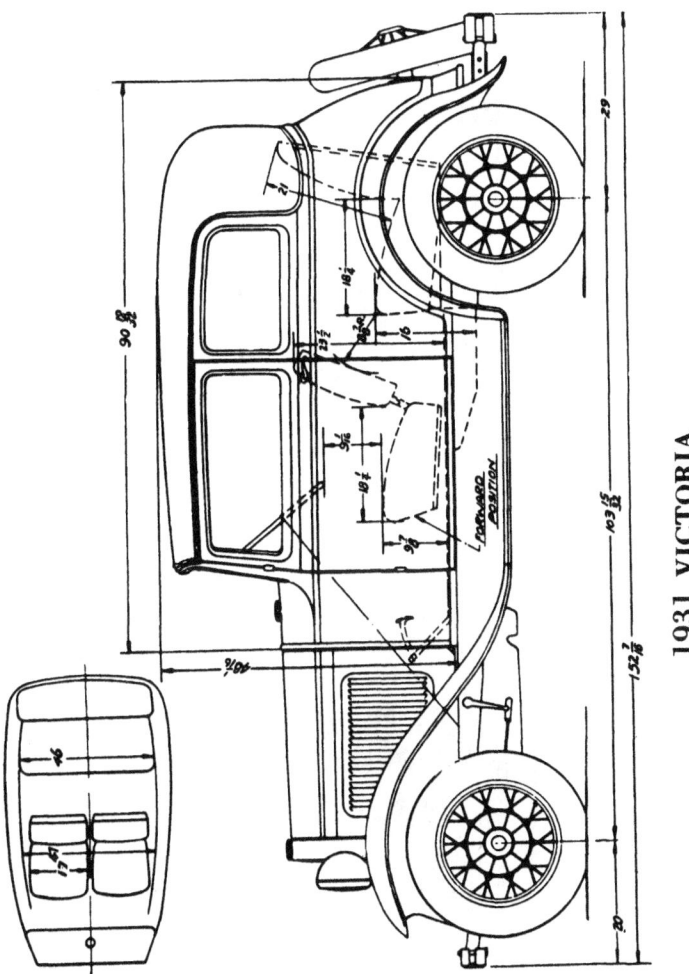

1931 VICTORIA

(Dimensions approximately same for De Luxe Phaeton, except it is 3 inches longer from back of front seat)

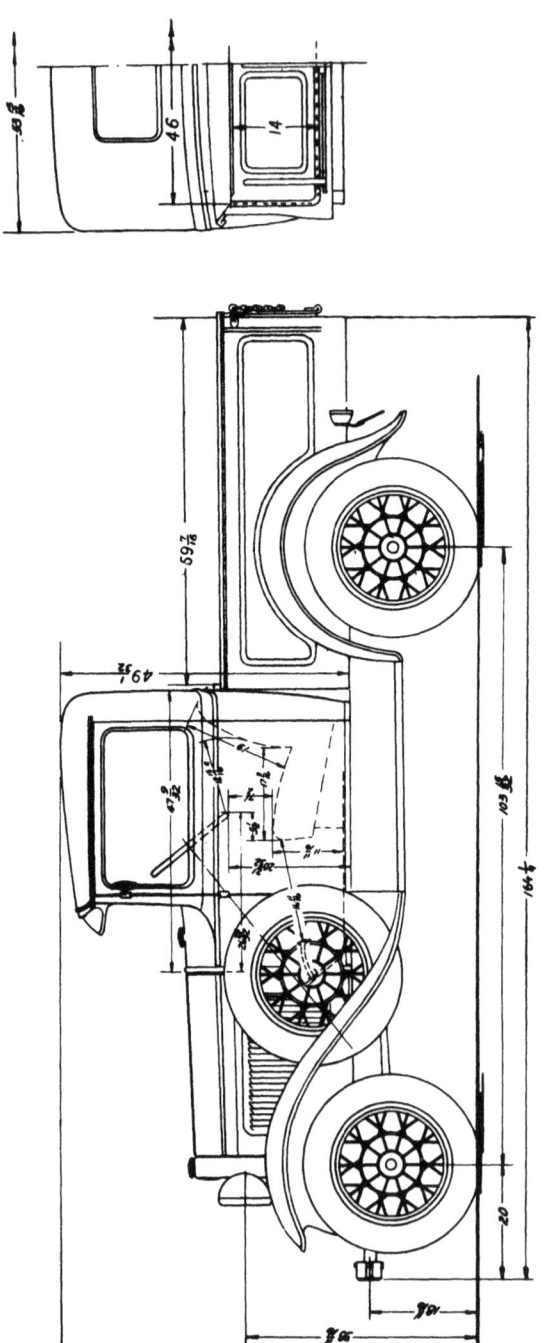

1931 CLOSED CAB PICK-UP

This dimensional drawing of the Closed Cab Pick-Up truck shows its construction on the standard MODEL A Ford chassis with the 103½" wheelbase.

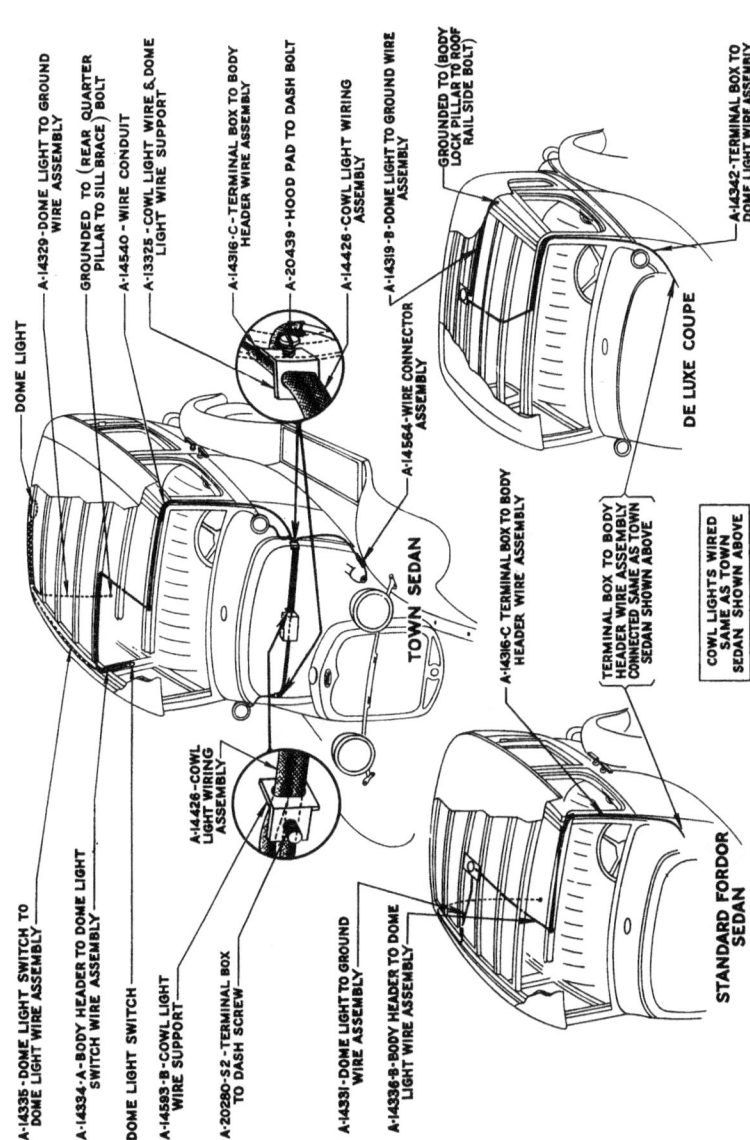

Fig. 938—Wiring Diagram of Town Sedan, Standard Fordor Sedan and DeLuxe Coupe

Ford Service Bulletin *for December*

Description and Installation Instructions of Victoria Coupe Top Material

The top material used on the new Victoria Coupe is a pyroxylin coated double texture fabric having an attractive fabric grain and printed pattern.

Covering the side and back quarters with this quality fabric adds to the attractiveness of the body and gives the car the appearance of a custom built job.

In the manufacture of the Victoria Coupe top the two side quarters and the top are first cut and then sewed together ready for mounting on the body. When installed it has the appearance of a one piece job.

So that mechanics will be thoroughly familiar with how this new top is installed, we are describing it below:

INSTALLING NEW TOP DECK ON VICTORIA COUPE

Care must be used when removing the old deck to prevent any possibility of marring the body finish. Be sure to remove all old tacks.

To install the new top deck, place the deck over the back of the body, sliding it forward towards front of car. The rear end of the deck is attached first. To locate the correct position at which to attach the new deck to the body, first locate the center of the rear window at both top and bottom of window. Next measure $21\frac{1}{4}''$ to both the right and left of the center line at bottom of window, and mark the body with a piece of chalk. (See Fig. 1050.) At the top of the window measure off $22\frac{1}{4}''$ and mark the body in the same manner. (See Fig. 1050.) The locating points at the top of the window are $1''$ further apart than at the bottom, due to the radius at the upper corner of the body.

Fig. 1050

Both bottom corners of top deck are next attached, using two 12 oz. tacks at each corner, and lining up the curtain with the markings. Make certain that both tacks go through the 3 layers of cloth at the seam so as to stand the pull when stretching top into place. All tacks must be placed in the perforated grooves so that the binding will cover the tacks. (See Fig. 1050.) Next tack the upper end of back curtain, using four 12 oz. tacks at each side. (See Fig. 1051.) Then starting at outer seam, slit the corner of the back curtain to within approximately $1\frac{1}{2}''$ of end of groove. (See Fig. 1051.) This prevents tearing when stretching curtain towards front of body. Trim off surplus stock close to the tacks at top of back curtain. This must be done before tacking down rear corner of top deck. Pull rear corner of top deck downward

Fig. 1051

so there will be approximately $1\frac{1}{4}''$ lap over and the seams in the upper deck line up exactly with the seams in the rear curtain. (See Fig. 1052.)

After tacking down both rear corners of upper deck, locate center line at front end of

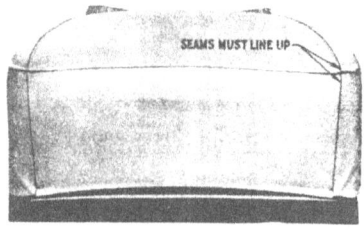

Fig. 1052

body and measure 20½" on each side of center line and mark body at that point. (See Fig. 1053.) Next line up the seams in the top deck with these markings and pull forward on both front corners (one corner at a time) until all wrinkles are smoothed out, then tack the corners down with two 12 oz. tacks. The seams must follow the contour of the roof.

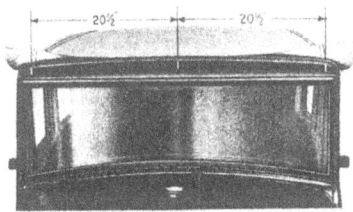

Fig. 1053

Pull rear quarter down as tightly as possible, making certain that all wrinkles are removed, then tack down with six 4 oz. tacks

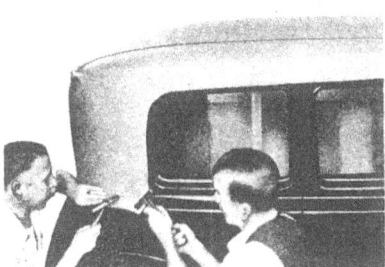

Fig. 1054

at outer corners. (See Fig. 1054.) Both rear quarters are tacked down in this manner.

Stretch sides and pull top deck forward and downward until all wrinkles disappear. The

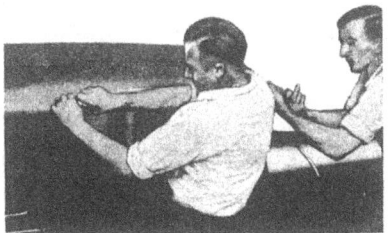

Fig. 1055

top deck must be drawn down very tightly when performing this operation. (See Fig. 1055.) Then tack down with four 6 oz. tacks along

Fig. 1056

edge of deck. Both sides must be tacked in this manner. Next tack all the way around the outer edge. The lower edge of the back curtain is then tacked down, also the lower edge of the slit section of back curtain.

Fig. 1057

Next tack down the upper edge in the same way. (See Fig. 1056.) Tacks must be very close together at these corners. (See Fig. 1057.)

Next pull upper edge of back curtain very tight until all wrinkles have disappeared, constantly pulling forward while installing each tack, then trim off surplus stock as shown in Fig. 1057.

Trim off surplus stock all the way around the tacked section, trimming as closely to the tacks as possible. (See Fig. 1058.) Next tack down the front end of the top deck, starting at the center and working outward. (See Fig. 1059.) Do not trim off surplus stock at front end until after moulding is installed.

Fig. 1058

Stretch rear end of top deck towards rear and tack down as shown in Fig. 1060, then trim off surplus stock close to the edge of the tacks.

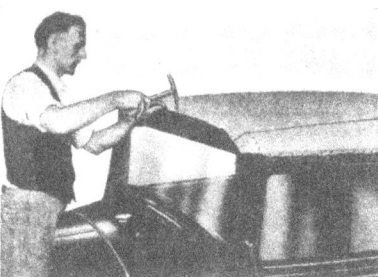

Fig. 1059

The back window is next cut out, leaving about 1½" of stock all the way around the edge of the back window. A sharp knife is used for performing this operation. (See Figs. 1061 and 1062.)

Next slit each corner of back window. At

Fig. 1060

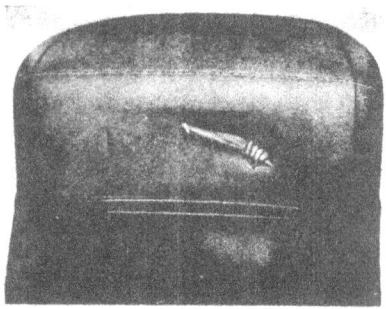

Fig. 1061

the corners the slits should not be over ¼" apart and as a starter they should not be over 1" deep. (See Fig. 1063.) This operation is very important. If not correctly performed, it will necessitate replacement of the deck. If necessary to lengthen the slot when tacking down,

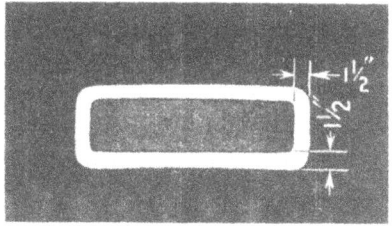

Fig. 1062

use a knife. These slits must not show from the outside after back window is tacked in. Figure 1064 shows back window tacked in place.

The binding is next installed. Due to the possibility of striking the head of the tacks

Fig. 1063

Fig. 1064

underneath the top material when tacking on the binding, time will be saved by first punching the holes for the binding. (See Fig. 1065.) Next install upper binding, using 12 oz. tacks. Position binding over tack line where top deck and rear curtain are tacked together, lining up

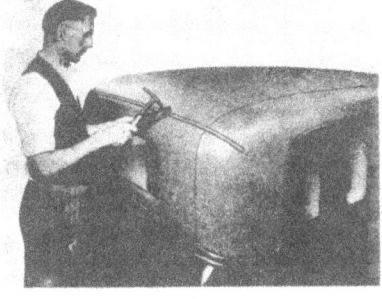

Fig. 1067

over and press it down tightly until a smooth surface is obtained. A hammer and fibre or hardwood block are used for this purpose. (See Fig. 1067.) Binding clip A-192140 is then placed at each end of binding. (See Fig. 1068.)

Fig. 1065

lower seam on binding with tacks already installed. Position the binding about ¼" back from end of groove. (See Fig. 1066.) Place several tacks in binding, then stretch tight and tack opposite sides of binding as shown in Fig. 1066. Next fold the binding

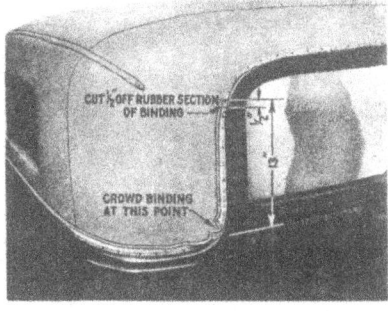

Fig. 1068

Next install lower binding. Locate correct position—measure 13" from upper moulding. (See Fig. 1068.) Then peel back binding and cut off ½" of the rubber section of the binding to allow end of drip moulding to fit down snugly

Fig. 1066

Fig. 1069

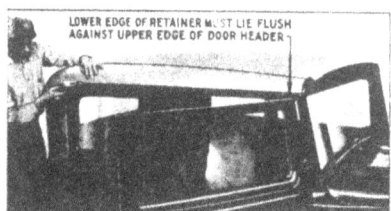

Fig. 1070

over the end of the binding. Start tacking at the 13" mark, pulling the binding tight and working downward. When you get down to the curve in the body, crowd the binding to allow it to wrinkle as shown in Figure 1068. Unless this is done the binding will not lie

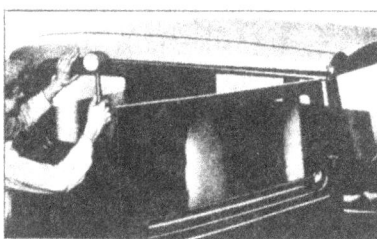

Fig. 1071

smooth when folded back. Next tack binding around to rear curtain seam, then stretch over to the opposite side and tack in place as previously described. After the tacking operation is completed, fold the binding back and flatten down, using a hammer and fiber block

Fig. 1072

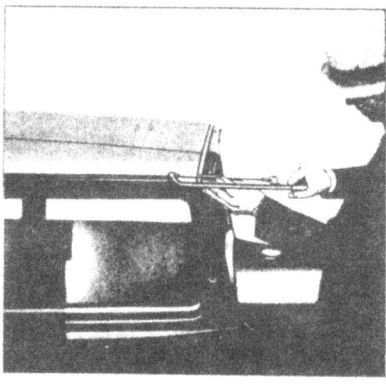

Fig. 1073

in the same manner as shown in Fig. 1067. Use edge of fibre or wooden block at corners where binding is tacked.

Figure 1069 shows upper and lower binding installed.

Next install drip moulding retainer. Place moulding retainer in position on 12½" mark on binding. (See Fig. 1068.) See that lower edge of retainer is flush against the upper edge of door header. Then nail retainer in place,

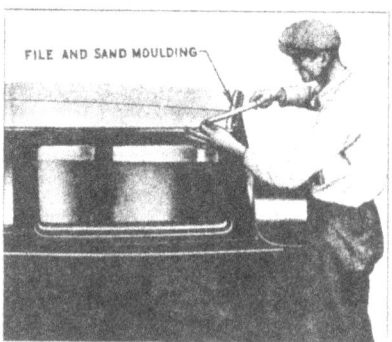

Fig. 1074

using 1" flat head nails and spacing the nails every fifth hole, then go over each nail with a nail set. *This is important.* (See Fig. 1070.) After installing drip moulding retainer on both sides of body, install drip moulding. This moulding snaps over the retainer. A rubber mallet is used for installing. (See Fig. 1071.)

After installing moulding, place the ¾" screw (removed from the old job) at front end

Fig. 1075

Fig. 1077

of moulding, and a 1" screw at rear end of moulding. (See Fig. 1072.) Next install front roof moulding, placing moulding over tack line at front of deck, and making certain that both ends of the front moulding touch the drip moulding. Start nailing the front moulding at center, and work outward so that moulding can be lined up as the nails are driven in. The nails must be set all the way down in the groove. A flat pointed punch can be used for this purpose. After nailing down moulding, pull the edges of the moulding together so that the nails do not show. The fibre or wooden block and hammer are used for this purpose as shown in Fig. 1067.

Cut out ends of roof moulding about $\frac{1}{16}''$ above drip moulding to prevent any possibility of squeak a hack saw is used for this purpose. (See Fig. 1073.) The moulding should then be filed down and sanded from seam in the top deck down to the drip moulding. (See Fig. 1074.)

Next trim off the surplus top deck material as close to the moulding as possible. (See Fig. 1075.) Use a sharp knife to prevent ragged edges. Next seal the edges of the front roof moulding from seam to seam (see Fig. 1076), also seal rear upper binding in the same way. (See Fig. 1077.) The sealing fluid in seam should be about $\frac{1}{16}''$ wide. A special sealing fluid is used for this purpose. The fluid can be obtained direct from the Dolphin Paint and Varnish Co., Toledo, Ohio. It is put up in both large and small cans. Fig. 1078 shows the completed job.

Fig. 1078

CLEANING VICTORIA COUPE TOP MATERIAL

A good grade of saddle soap or ivory soap worked into a lather and applied with a sponge or soft cloth will clean the majority of these tops satisfactorily. Carbon tetra chloride can also be used with very good results if care is used in applying it. To use it, moisten a soft cloth with the chloride, then rub the soiled spot very lightly. Do not apply too much pressure or it will remove the grain from the material.

Fig. 1076

PARTS SUPPLIERS AND SOURCES

While the Ford Motor Company and the Ford dealers have *not* maintained stocks of MODEL A parts for many years, it is possible that your local Ford dealer *may* have a few parts still remaining on his shelves, and nearly every Ford dealer can arrange to supply a rebuilt MODEL A engine on order.

Many individuals have acquired parts which they offer for sale by mail order as a part of their antique automobile hobby activity. Some few of these have built this activity up to the proportions of small business enterprises and publish periodic lists of available parts. Others simply specialize in certain parts or accessories, some of which are being newly manufactured to meet the increasing demands of MODEL A enthusiasts.

MODEL A parts may also be located through the classified advertisement sections of the publications of many Ford and antique automobile clubs in America.

Several of the major mail order houses still list MODEL A (and Model T) Ford parts in their catalogs.

Here, without recommendation of any kind, are alphabetical listings of the several major club publications, the many individuals, and the three major mail order houses offering MODEL A Ford parts and accessories:

1. CLUB PUBLICATIONS

These are not the only clubs devoted to the automotive hobby, but these are large clubs publishing regular magazines, and offering many membership benefits:

The ANTIQUE AUTOMOBILE
 published bi-monthly by:
 The Antique Automobile Club of America, Inc.*
 Hershey Museum, Hershey, Penna.

The BULB HORN
 published quarterly by:
 The Veteran Motor Car Club of America
 15 Newton Street, Brookline 46, Mass.

The HORSELESS CARRIAGE GAZETTE
 published bi-monthly by:
 The Horseless Carriage Club of America, Inc.
 7730 South Western Ave., Los Angeles 47, Calif.

The MARC NEWS
   published monthly by:
   The Model "A" Restorer's Club
   P. O. Box 615, Zanesville, Ohio
The RESTORER
   published quarterly by:
   The Model A Ford Club of America, Inc.
   6924 San Fernando Road, Glendale, Calif.

*This club caters to all types of antique, classic, and production cars and has a division for Model A Fords.

2. INDIVIDUAL SUPPLIERS

When writing to a supplier be sure to list specifically the parts you want and identify them by Ford number, if possible. Remember that this is usually a hobby on the parts of suppliers, so be sure to enclose a stamped, self-addressed envelope for a reply. Remember, too, that existing supplies of these parts are limited; do not be surprised if some of the parts you want are gone by the time you write!

AETNA RUBBER COMPANY
108 Broad Street
Boston 10, Mass.
   Tops & curtains ready-made

O. R. ALLMAN
285 DeSmet Drive
Florisant, Missouri
   Parts list available. Has new back & seat cushions for A & T Fords.

GEORGE DE ANGELIS
9830 Allen Road
Allen Park, Michigan
   Duplicate FORD patent date plates & ignition switch plates only.

ANTIQUE ACCESSORY SHOP
2125 Lacrosse Avenue
St. Paul 6, Minn.
   Flying Quail mascot for radiator cap only.

ATLANTIC AUTO & TRUCK SALES
19020 South Figueroa
Gardena, Calif.
   Used parts for all makes of cars.

BARRETT & SONS, INC.
Main Street
Matawan, New Jersey

WILLIAM F. BEAVER
2724 Pasteur Avenue
Overland 14; Missouri
  Large Stock. Parts list available.

BEN'S AUTO WRECKING
10225 Glenoaks Boulevard
Pacoima, Calif.
  Large Stock. Parts list available.

CORNELIUS BERBOWER
Rt. #3
Newton, Illinois

DON BESTLAND
122 East Laurel Street
Glendale 5, Calif.
  Many new parts.

BILL BITTMANN
4154 North Drake
Chicago, Illinois
  Large Stock. Parts list available.

BRUER AUTO PARTS & REBUILDERS
1900 Blaine Street
Springfield, Missouri

ROBERT C. BURCHILL
2316 17th Avenue
Port Huron, Mich.
  Parts Catalog, 50c.

WALT CANN
324 North Fullerton Avenue
Montclair, New Jersey.

COMPETITION ACCESSORIES
704 Washington Ave., Box 141
Iowa Falls, Iowa.
  New Moto-Meters; Clock mirrors.

ROBERT J. CARINI
67 Denslow Road
Glastonbury, Conn.
  Send your want list.

G. C. CHEATHAM
708 Robinson Drive
Birmingham 6, Alabama.

KENNETH COULTER
Sevierville Road
Marysville, Tenn.
  Send your want list.

JOE CUZELIS
20724 Fairview Drive
Dearborn, Mich.

DRAWBRIDGE AUTO SUPPLY
Broadkiln River, Rt. 14
Milton, Delaware.
   Send your want list.
EDWARD DUDIK
5810 - 44th Avenue
Hyattsville, Maryland
   Send your want list.
EGGE MACHINE COMPANY
(Nels Egge)
7704 South Main Street
Los Angeles 3, Calif.
   Manufacturer of aluminum pistons.
ELMER'S AUTO PARTS
Webster, New York.
   Send your want list.
BILLY VON ESSER
3307 West Irving Park
Chicago 18, Illinois.
BOB FIGGE
1916 North Wilson
Hollywood, Calif.
DENNY FEATHER
P.O. Box 96
Bruceton Mills, West Va.
   Send your want list.
FORD PARTS OBSOLETE
616 East Florence
Los Angeles 1, Calif.
   Send your want list.
GAYLE AUTO PARTS
1019 McKee Street
Houston, Texas
SHORTY GIBSON'S
Rt. #8
Marysville, Tenn.
   Large Stock. Send your want list.
CHUCK HAFNER
Speedometer Service Company
4740 Baum Boulevard
Pittsburgh 13, Pa.
   Speedometers and parts.
BILL HALL
P.O. Box 615
Zanesville, Ohio
FRANK HANKINS
Route 25, R.D.
Riverside, N.J.
JOHN HANSEN
350 Paderewski Avenue
Perth Amboy, N.J.

J. L. HELMS, JR.
3912 Plaza Street
Charlotte, N.C.
   Manufactures new running boards, trim, etc.

E. R. HEMMINGS
P.O. Box 433A
Quincy, Illinois.
   Large Stock. Parts list available.

RAY HOVE
5606 Clinton Avenue
Minneapolis, Minn.

RICHARD G. HUBER
See MARK Auto Co.

JIM'S SOHIO SERVICE
1131 Cleveland Avenue
Ashland, Ohio.

JUDSON MANUFACTURING COMPANY, INC.
Cornwells Heights, Penn.
   Manufacturer of aluminum pistons.

DONALD KAUFMAN
97 Appleton Avenue
Pittsfield, Mass.

NORMAN KAYE
45 Chamberlin Drive
Buffalo 10, N.Y.

DALE KILBORNE
4703 Hersholt Avenue
Long Beach 8, Calif.
   Aluminum FORD runningboard step plates.

CHARLES KLINGER
114 South Mountain Avenue
Montclair, N.J.

LEO'S MODEL T REPAIR
911 East 18th Street
Kansas City 8, Missouri.
   Also supplies MODEL A parts.

RICHARD LA SALLE
(See MODEL A FORD parts & service).

J. S. MAPLES
4122 - 5th Avenue, South
Birmingham, Alabama.
   Send your want list.

MARK AUTO COMPANY, INC.
Layton, New Jersey

PAUL MARVEL
62 Spencer Avenue
Lancaster, Pa.

MIDWEST CLASSIC & ANTIQUE CARS & PARTS
(Ray Butler & Bobb Tapp)
1837 California
Denver 2, Colorado
  Send your want list.

JOE McCLELLAND
See FORD PARTS OBSOLETE

McKENQUIE MOTOR SALES
109 Broadway Street
Cambridge, Mass.
  Send your want list.

MODEL A FORD PARTS & SERVICE
(Richard La Salle)
907 Sterner Mill Road
Trevose, Bucks Co., Pa.
  Send your want list.

MURCHIO'S MOTOR CAR MUSEUM
(Joseph J. Murchio)
Greenwood Lake, N.Y.

JOE OSTERMAN
Box 234
Sugarcreek, Ohio
  Original type mufflers.

PARKINSON MOTOR SALES
Ponca City, Oklahoma.
  Send your want list.

R. N. PEDRICK
131 Kent Avenue
Kentfield, Calif.
  Original type motometers.

GENE RENNINGER
123 Lincoln Street
Lancaster, Pa.
  Large Stock.

ED. P. RYAN
530 Broadway
Malden 48, Mass.
  Send your want list.

G. S. SCHULZE
1207 South 91st
West Allis, Wis.
  Send your want list.

REID SHAW
2009 Edgewood Avenue
High Point, N.C.
  Send your want list.

K. L. SLINGERLAND
Cherry Ridge Farms
Westtown, N.Y.

HARRY R. SMITH
81-35 Margaret Place
Glendale, N.Y.

OLLIE L. SMITH
304 Jerome Avenue
Linthicum, Maryland

DALE STOCKMAN
R & F Mfg. Co.
Box 166
Bourbon, Indiana.
   Original type wind wing brackets.

AL VIVIAN ANTIQUE AUTO PARTS
156 - J Street
San Bernardino, Calif.
   Send your want list.

WARSHAWSKY & COMPANY
1900 - 24 South State Street
Chicago 16, Illinois.
   Original type mufflers.

CHARLES WENDLING
Frankfort, Kansas

J. R. WEATHERLY, JR.
Rt. #2, Box 5
Crossett, Arkansas

B. S. WISNIEWSKI, INC.
201-245 West Maple St.
Milwaukee 4, Wisconsin

JAMES A. WOLFRAM
2845 - 40th Avenue
Minneapolis 6, Minn.
   Chassis & body parts; send your want list.

ED. WRIGHT
876 Bay Street
Springfield, Mass.
   Send your want list.

DESCRIPTION OF THE FORD MODEL "A" CARS

The new Ford is a car that conforms more to standard practice than the Model T, and the well-known features of that car, the flywheel magneto and the planetary transmission, are not found in the new model, which has a standard gear shift and a conventional electrical system. The body lines of the new cars are a marked improvement over those found in the old car. The side view of the Tudor sedan, a very popular model, shown at Fig. 1, is typical of the graceful proportions of the entire line. The front view of the phaeton, shown at Fig. 2, shows the new radiator design and lamp hook-up.

The Model A has a four-cylinder engine of $3\frac{7}{8}$ in. bore and $4\frac{1}{4}$ in. stroke, which gives a piston displacement of 200.5 cu. in. The engine is oiled by the circulating splash system, the pump delivering into the valve

The Ford Model A Car

chamber, from which the oil flows by gravity to each of the main bearings. Connecting rod heads are provided with dippers which splash oil from troughs in the lower part of the crankcase to all of the other parts of the engine requiring lubrication. The transmission, which is of the regular sliding pinion type, has gears made of

Fig. 1.—Side View of Ford Model A Tudor Sedan Showing Graceful Body Lines and Modern Appearance.

chrome steel. Four-point suspension is used for the engine, the rear support brackets on the flywheel housing bolting directly to the frame side members, thus forming an additional cross-member, while at the front the engine is supported on a rearward extension, in the form of an inverted channel, of the frame front cross-member. As has been mentioned, the clutch and transmission are similar in design to the corresponding units used in the Lincoln, the transmission being of the three-speed, standard-shift type, and the clutch a multiple disk, dry design.

Details of Model A Engine

Drive to the rear axle is by an inclosed propeller shaft with ball-joint at the forward end, to the propeller-shaft-tube member of which the diagonal radius rods from the rear axle housing are secured. The rear axle is of three-quarter floating type. Both the generator and the starting motor are located on the left side of the engine (looked at from the rear), the former being driven by the V-type fan belt, while the latter drives to the flywheel. The attention of the reader is directed to the

Fig. 2.—Front View of Ford Model A Phaeton Showing Radiator Design, Full Crown Fenders and Lamp Installation.

The Ford Model A Car

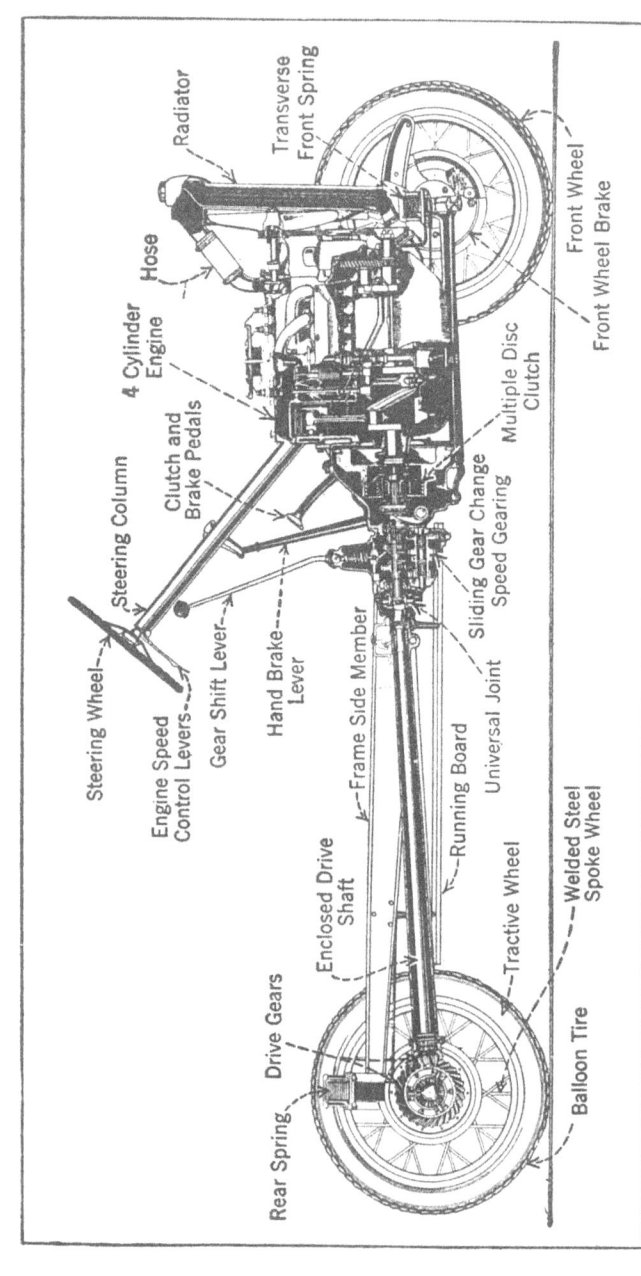

Fig. 3.—Sectional View of Ford Model A Chassis Showing Power Plant in Part Section, also Clutch, Three Speed Sliding Gearset and Shaft Drive to Rear Axle.

Details of Model A Chassis

sectional view of the chassis shown at Fig. 3, as this clearly shows the main details of the chassis. As can be clearly seen, three cross-members are used in the frame. These are all of the gusseted type, the front member having the form of a double channel, while the others are of the regular channel type. In contrast to statements recently made in various automotive publications, the frame members are riveted together and not welded. The cross-members, of course, are supplemented by the rear engine supports, which virtually form an additional cross-member. There is also a pressed steel member tying together the rear bumpers. An outward bulge of the rear part of the frame is also noticeable.

Radius rods for the front axle have also been redesigned. Like those for the rear axle, they are of the tubular type with an elliptic cross-section, and they connect at the rear end to the bottom of the transmission case through a ball joint. At the front end, they connect to the axle by means of yoke-and-pin joints, the pins passing vertically through eyes in the yoke and the front axle. Wire wheels of the same type as used on the last Model T series are standard equipment, a spare wheel being included. These wheels take 29 x 4.50 in. balloon tires.

The Model A Ford Bodies.— Bodies are of composite construction. The body and pillars, and the roof side rails are of wood, sheet-steel-clad, and are covered with sheet-steel panels. On the closed models complete body side panels are built up and flash-welded onto the rear body panel. Tops are fabric-covered. Cowls are made in three pieces, the top part being formed by the gasoline tank, to which the side panels are flash-welded. Fenders

The Ford Model A Car

are full-crowned, belt lines are higher, and the wheelbase is longer.

A new instrument panel is adopted. This is roughly of diamond-shape, with the dash gasoline gage, reading in quarters of tank capacity, at the top, the speedometer at the bottom, the ammeter at the right, the ignition lock at the left, and the instrument board lamp in the center. The ignition lock is of the Electrolock type, grounding the distributor, the connecting cable being enclosed in a metal conduit. Spark and throttle control levers are still mounted on the steering column below the wheel, but the former quadrants have been eliminated, and a foot accelerator is provided. Above the wheel on the steering column is mounted the light control lever. Double-filament, double-contact bulbs are used in the bullet-type headlamps.

A combination tail and stop light is standard equipment, as are also Houdaille shock absorbers, front bumpers and rear bumperettes. The latter attach to both the frame and the body and are curved to carry out body lines. They, as well as the front bumpers, are assembled on the cars before the body is put on, but are easily removable. Additional items of equipment are an automatic windshield wiper, a swinging type one-piece windshield with ventilating ports in the top of the cowl exposed when opening the windshield, a speedometer, which is driven from the front end of the propeller shaft, and a Sparton horn which is mounted between the radiator shell and the left fender below the headlamp. The crank handle is removable, whereas in the Model T it could not be removed. The running boards are of steel construction, rubber-covered, with aluminum bindings and concealed screws.

Equipment of Model A Cars

Interior finish of the cars is also decidedly improved. On the open models and the cabriolet, unpleated brown imitation leather is used for the upholstery, the seat cushions having a single pleat running lengthwise of the cushion, about two-thirds of the way forward. In the closed cars cloth upholstery is used, with doors and side walls trimmed in the same materials, the doors being fitted with pockets also made of pleated upholstery material. Door controls are of the remote type, operating by lifting the lever, rather than by turning it. Windows are crank-operated, and door snap locks are provided inside the closed cars. On the cabriolet a regulation roll-up type of rear curtain is provided. A choice of four colors is being offered, namely, Niagara blue, Arabian sand, Dawn gray, and Gunmetal blue. The belt and reveals are in all cases in contrasting colors and the bodies are attractively striped. All cars are finished in pyroxylin lacquer. The fenders are of the full-crowned type and harmonize with the body lines.

Seat cushions are deeper and softer than those of the Model T. Door handles and window lifts are nickel-plated. The speedometer, gasoline gauge, a meter and ignition lock are mounted on an instrument panel of satin-finish nickel and are illuminated by a lamp mounted in the center. The headlamps and radiator shell are nickel-plated on all exposed surfaces. The closed cars have the cadet type of sun visor and a crown roof. Unusually narrow front pillars, together with steel body construction, assure clear vision without impairing body strength.

As will be seen in the mechanical descriptions to follow, the car should be quiet. The body design works to that end, and the very flexible spring construction, work-

The Ford Model A Car

ing in conjunction with hydraulic shock absorbers, minimizes road shocks, thus lessening body noises. Care has been exercised in the design of the bodies to prevent squeaks, rattles and drumming sounds.

The Ford Model A Engine.—The engine of the Model A is of brand-new design, and none of its major parts apparently are interchangeable with those of the Model T. For instance, the crankshaft, while still of the three-bearing type, is counterbalanced by machining the first, third, fourth and sixth crankcheeks in cylindrical form.

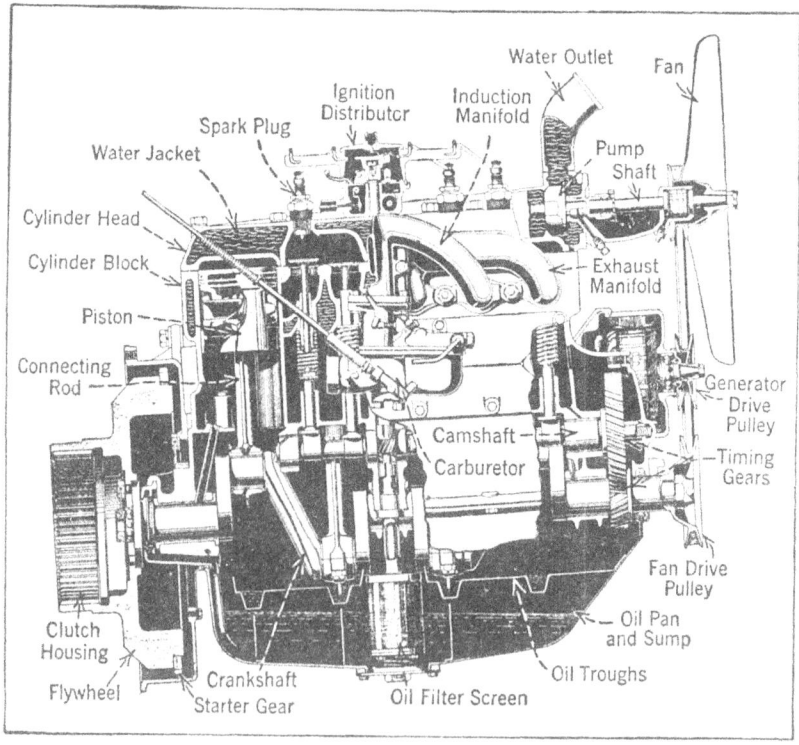

Fig. 4.—Part Sectional View of Ford Model A Engine Showing Design of Important Internal Parts.

The Ford Model A Engine

Connecting rods are longer and lighter. The sectional view of the engine at Fig. 4 shows the sturdy and simple mechanical construction of the engine very clearly, while the various important components of the power plant shown in illustrations on Fig. 5 serve to make the design of these parts clearer to the reader. The power and accelerating qualities of the engine are remarkable for a low-priced car. In tests in high gear, it is stated that a Model A with a two-door sedan body, with two passengers, is said to have accelerated from 5 to 25 m.p.h. in $8\frac{1}{2}$ seconds, and the maximum road speed varies between 55 and 65 m.p.h. Some road tests have shown a speed in excess of 65 m.p.h., but, quite naturally, the maximum speed which can be attained varies with road conditions, the style of body and the load carried. At 2200 r.p.m. the four-cylinder engine develops 40 hp. on the brake. While the rate of gasoline consumption will vary widely with service conditions, the car will make between 20 and 30 miles per gallon.

A turbulence type of cylinder head is now used and a higher compression ratio provided for. Spark plugs are mounted in the center of the combustion space, and are no longer of the $\frac{1}{2}$ in. type but are $\frac{7}{8}$ in. in diameter with 18 threads per inch, as in other makes of cars. Camshafts are of the three-bearing type, but are new in design, and have the drive gear for the vertical accessories shaft machined at the center. Of course, the engine is still of the L-head type. Camshaft drive is by a non-metallic gear, with no idler. The smooth operation and unusual accelerating qualities of the car are due, to a great extent, to the use of aluminum pistons. A three-bearing crankshaft is used, which is provided with circular disks concentric with the shaft in the places of the

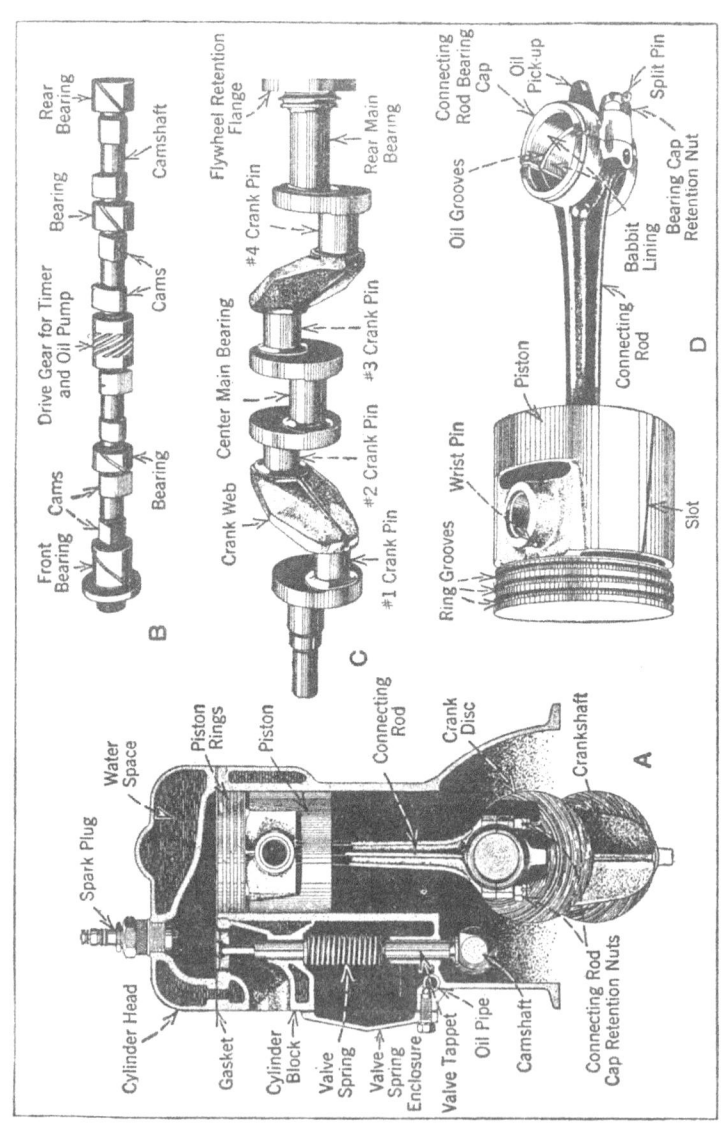

Fig. 5.—Diagrams Showing Important Structural Parts of Ford Model A Engine. A—Sectional View of Cylinder Showing Valve Action, Piston and Connecting Rod and Crankshaft and Their Relation. B—Camshaft Has Large Bearing Area. C—The Crankshaft Has Disc Webs and Is a Three Bearing Type. D—Piston and Connecting Rod Assembly.

Details of Lubrication and Cooling Systems

usual short arms. This shaft is statically and dynamically balanced, and as a result of the relatively low speed at which the full power is developed, the engine runs with exceptional smoothness. Cam contour and valve clearances have been so worked out that—with non-metallic timing gears—an unusually quiet power-plant results.

Ford Model A Lubrication and Cooling Systems. — Engine lubrication is unique and of distinctive Ford design. The system involves pump circulation, splash and gravity feed. Oil from the pump is delivered to the valve chamber, from which it flows by gravity to the main bearings of the crankshaft. There is an oil dipper on the cap of each connecting rod, which splashes the oil over the bearings and distributes it to all interior parts of the engine. This is shown in the sectional view of the engine on Fig. 6. Lubrication is by force fed to the main bearings, and by splash to all other parts. The oil pump is located at the lowest point of the pressed steel oil pan, a cover below it permitting of its removal without removing the oil pan. Oil is also delivered by pressure to the valve chamber on the right into the sump through a large external pipe.

Cooling is by a centrifugal water pump mounted on the cylinder head and combined with the fan. The cylinder head is provided with a riser to prevent the pocketing of steam in the rear end of the block. Water enters the block centrally at the left side. A two-bladed fan is used, driven by a V-belt, the belt also driving the generator, which is mounted forward on the left side of the engine. Adjustment is by swinging the generator to tighten the fan belt. The details of the cooling system, with all important parts designated, are shown in Fig. 7.

The Ford Model A Car

Carburetion and Ignition Systems.—Two of the auxiliaries which have a material influence on engine operation are the carburetion and ignition systems. The fuel

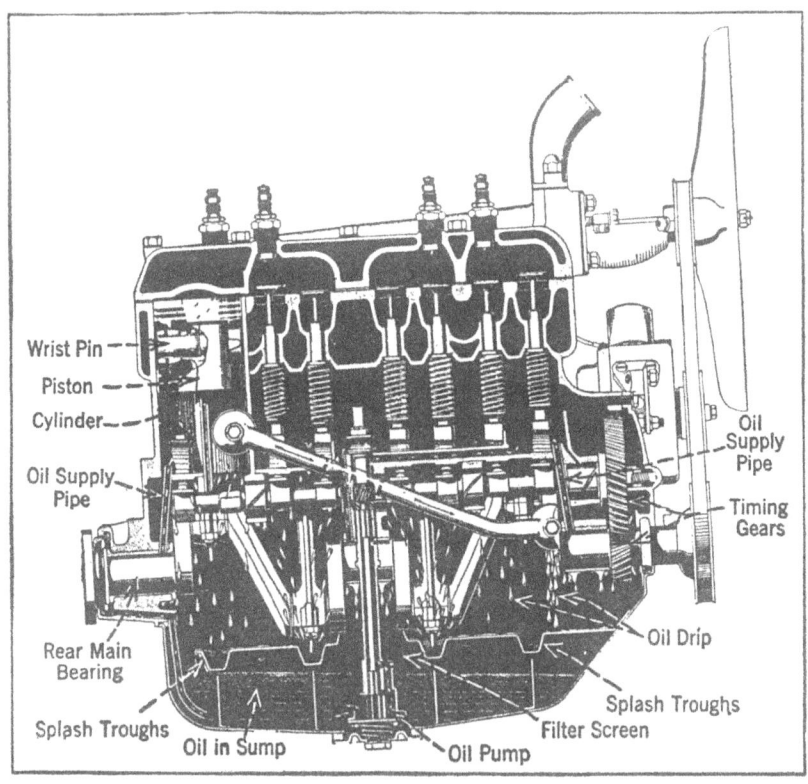

Fig. 6.—Sectional Side Elevation of Ford Model A Motor Showing Oil Pump and Method of Lubrication.

supply and carburetor of the Ford Model A engine are clearly shown at Fig. 8. The fuel tank is part of the body, being placed under the cowl, with the filler cap projecting. It includes a filter screen and fuel gauge. The capacity of the tank is 10 gallons. The carburetor is a

Details of Model A Cooling

Zenith 1" vertical, with choke and needle adjustment rod on dash. The hot spot ram's horn type induction manifold is clearly shown. The fuel system, with its gravity

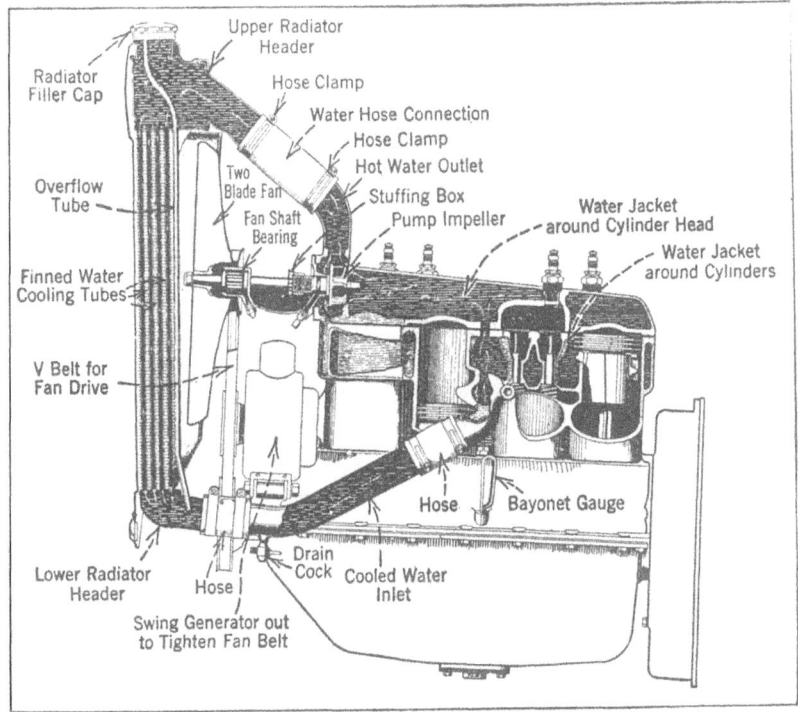

Fig. 7.—Sectional Side Elevation of Ford Model A Cooling Method Which Is a Combined Pump and Thermo-Syphon System.

feed and efficient carburetor, is comparable to that of the highest-grade cars.

The electrical system is illustrated in Fig. 9 and is of the usual standard design. The ignition system is of conventional modern design, the time-honored four coils and timer of the old Model T being done away with. One

The Ford Model A Car

ignition coil is used, and this is inclosed in a waterproof case. The distributor is located on the top of the engine, thus making it readily accessible and protecting it from

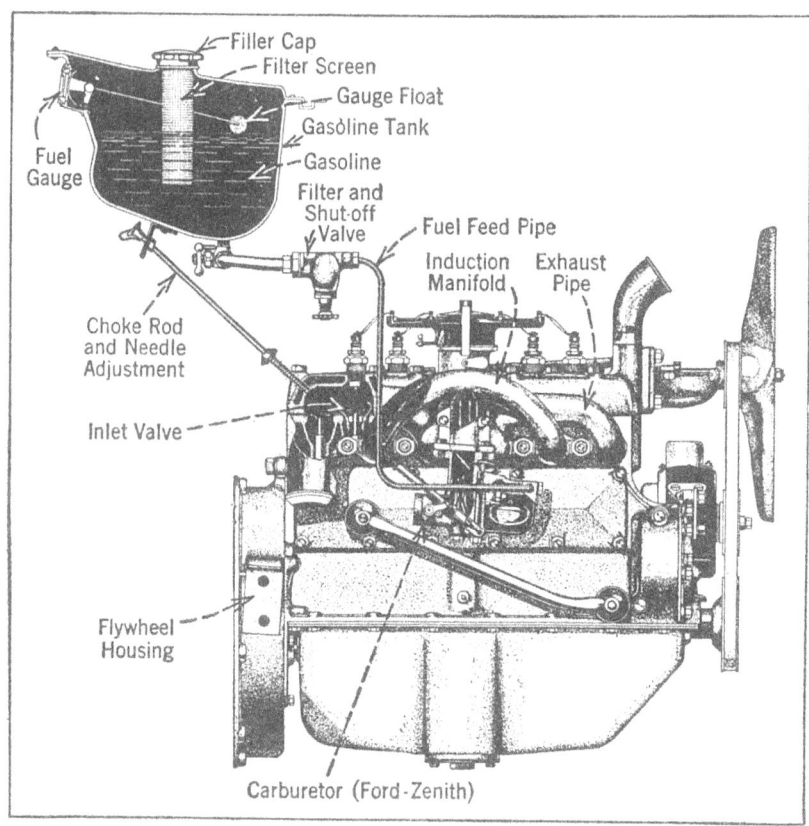

Fig. 8.—The Ford Model A Fuel System and Carburetor Installation.

oil and moisture. Connections from the distributor to the plugs are made by short bronze springs. A coincidental lock is placed in the ignition circuit. This not only replaces the regular ignition switch, but in the "off" position it grounds the entire circuit. A steel conduit

Details of Model A Electrical System

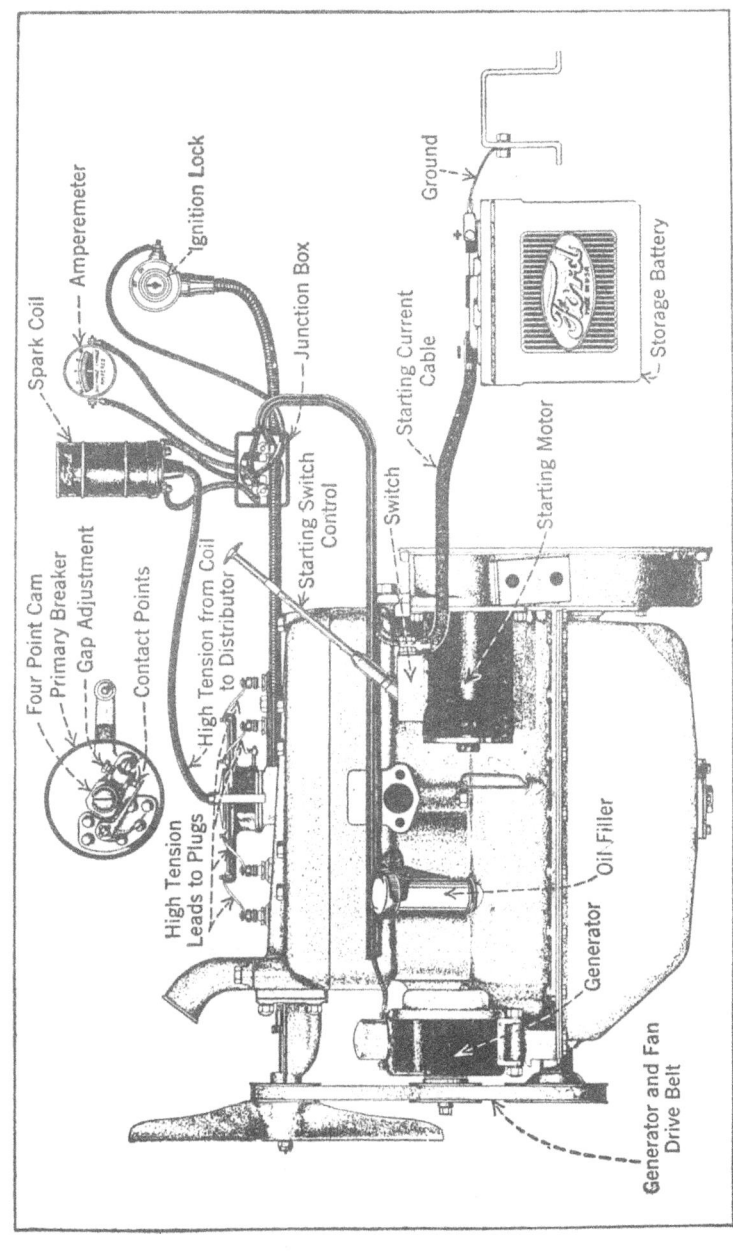

Fig. 9.—Ignition and Starting System of the Ford Model A Car. Note Construction and Location of Distinctive Ignition Distributor.

The Ford Model A Car

leading from the switch to the distributor protects the primary wire. This wire is grounded to the distributor casing, thereby making it impossible for anyone to wire around the device.

As has been mentioned, the generator, which is of Ford design, is driven by the fan belt. The starter is conventional and is mounted on the front of the flywheel housing on the side opposite to the generator. The ignition system is of American Bosch manufacture, but has the Ford name stamped on its units. The new distributor is mounted on top of the cylinder head, where it is driven by a vertical shaft, gear-driven from the center of the camshaft, the lower end of the vertical shaft driving the oil pump.

The cover for the new distributor is of molded composition, and is unique in that it has extensions running about 4 in. fore and aft over the cylinder block. Incased in these extensions are the leads for the various spark plugs. These leads are brought to the top of the extensions, where short brass strips are fastened down with round finger nuts. The other ends of these strips attach directly to the plugs. The coil for the new ignition system is mounted on the front of the dash.

Transmission System and Brakes.—The car has a three-speed-and-reverse selective sliding gear transmission with standard gear shift. The main shaft runs on ball bearings, while the gear cluster on the stationary secondary shaft is mounted on roller bearings, only the reverse idler having a plain bearing. This construction will at once be recognized as unusual in the lower-priced class. All gears are made of heat-treated chromium steel. In conjunction with this transmission, use is made of a

Model A Transmission System

multiple disk dry clutch, with four driving and five driven disks. This design is comparable with transmissions found in cars selling at ten times the price of the Ford Model A, and is shown in Fig. 10. The steering gear is

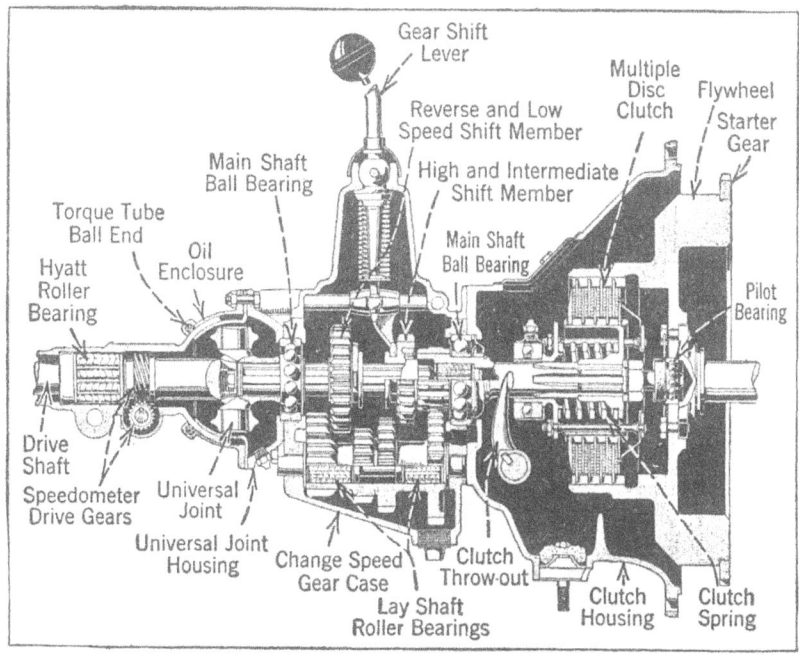

Fig. 10.—Sectional View of Ford Model A Multiple Disc Clutch and Three Speed Sliding Gear Transmission. Note Liberal Use of Anti-friction Bearings.

of the irreversible wheel and sector type and is shown at Fig. 11. The housing of the steering gear is of forged steel and has the tubular steering column welded to it. The ratio of the steering gear is 11¼ to 1. The steering wheel is of large diameter and is made entirely of steel, covered with hard rubber. The lighting switch and horn button are conveniently located on top of the wheel.

As formerly, torque tube drive is used. This torque

The Ford Model A Car

tube is tapered from about the middle of its length forward to the universal joint housing, and is bolted to the axle housing at the rear. Radius rods are of hollow elliptic section and connect by a ball joint to the torque

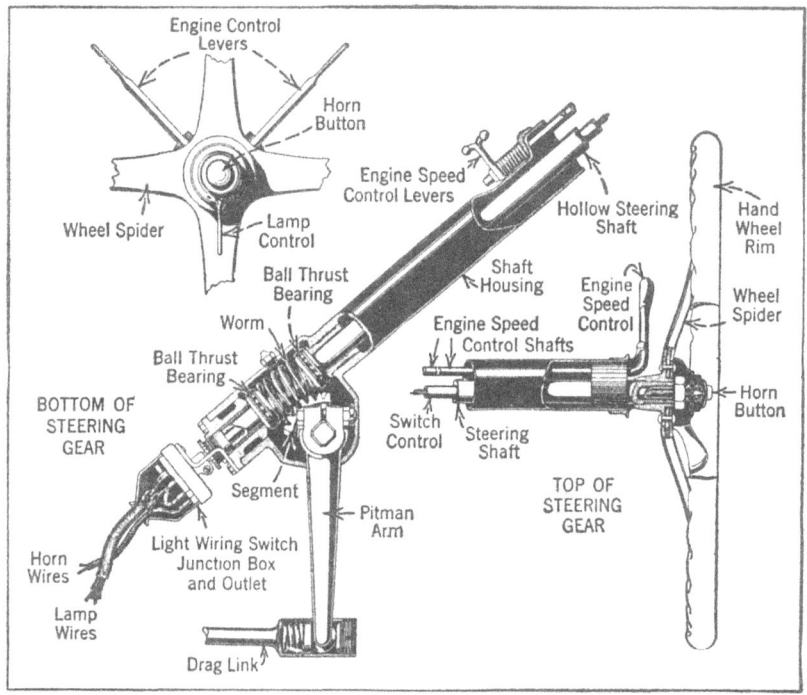

Fig. 11.—The Ford Model A Steering Gear is a Non-Reversible Worm and Segment Type.

tube immediately back of the transmission. A built-up rear axle housing is used, with the two sides of the housing bolted to the central cylindrical part inclosing the differential and drive gears. Two large, hollow-head plugs are used for filling with oil and draining respectively. Axle shafts are keyed to the wheel hubs. There

Model A Rear Axle

is no support for the axle shaft inside of the axle housing, but the load is taken by the axle housing itself from the wheel hubs directly, the latter being mounted on pilot roller bearings on case-hardened steel sleeves pressed onto the ends of the housing. The construction of the driving gear and differential assembly can be easily determined by examination of the illustrations in Fig. 12, while the rear wheel hub and bearing construction are easily determined by consulting the sectional view at Fig. 13.

A three-quarter floating rear axle is employed, and the axle shafts carry none of the weight of the car. The axle housings are made entirely of steel and are built up by welding steel forgings to steel tubing. The differential housing, to which these axle housings are bolted, is made of rolled channel steel. The wheels are carried on roller bearings on the axle housing, hence the shafts are not called upon to carry the weight of the car. All bearings in the rear axle are of the roller type, and the drive is by spiral bevel gears. The wheels, which are commonly referred to as wire wheels, are really steel spoked wheels. Each wheel is assembled by welding and becomes one piece of metal. This prevents the spokes from coming loose, and each spoke has a tensile strength of 4,000 lbs. As the outside spokes do not cross, and as there are only 30 spokes in each wheel, the wheels are particularly easy to clean.

Of interesting design are the mechanical four-wheel brakes. These are of the internal, two-shoe type and are self-centering. Smooth action and ease of adjustment are assured. All adjustments are made from outside, without removing any parts and without special tools. Both the brake pedal and the lever operate all four

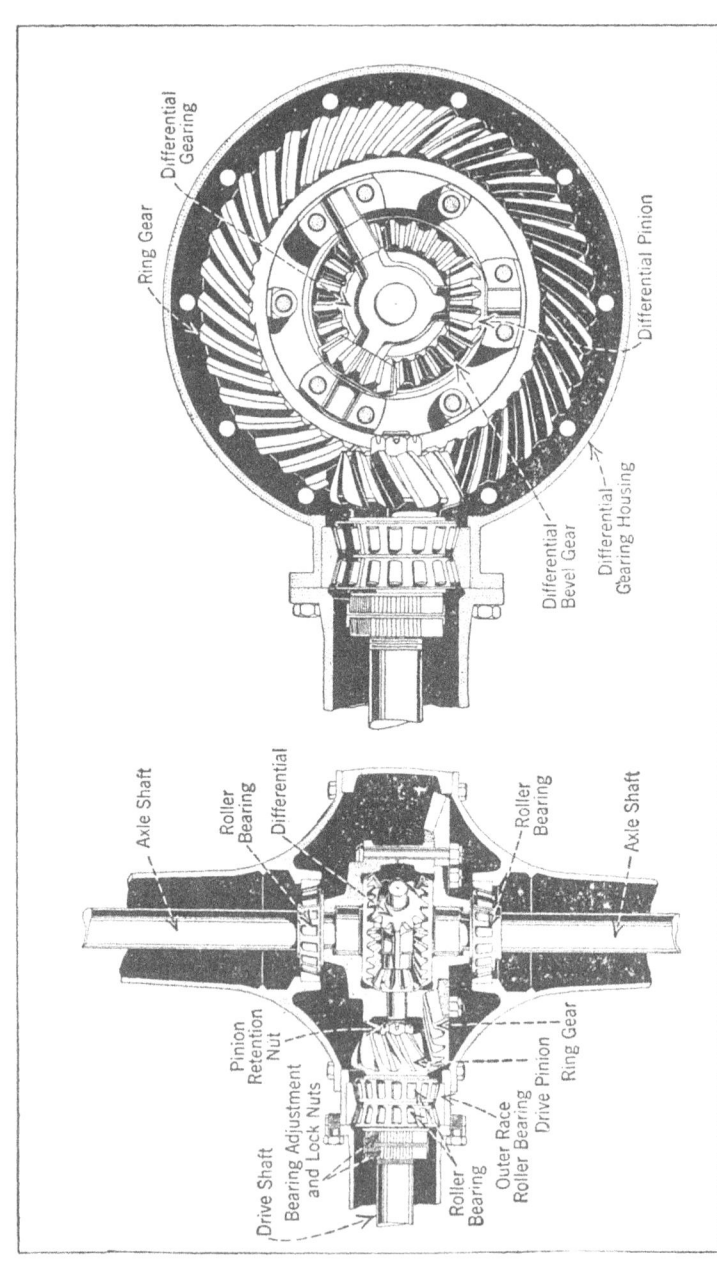

Fig. 12.—Sectional Views Outlining Construction of Bevel Driving and Differential Gearing of Ford Model A Rear Axle. Note Use of High Grade Taper Roller Bearings and Spiral Bevel Gears, Features Found in Higher Priced Cars. Plan Sectional View of Differential Housing Shown at Left, Side Sectional View at Right.

Details of Model A Rear Wheel

brakes. The total braking surface is 168 sq. in. In order to assure continued service and easy adjustment, all

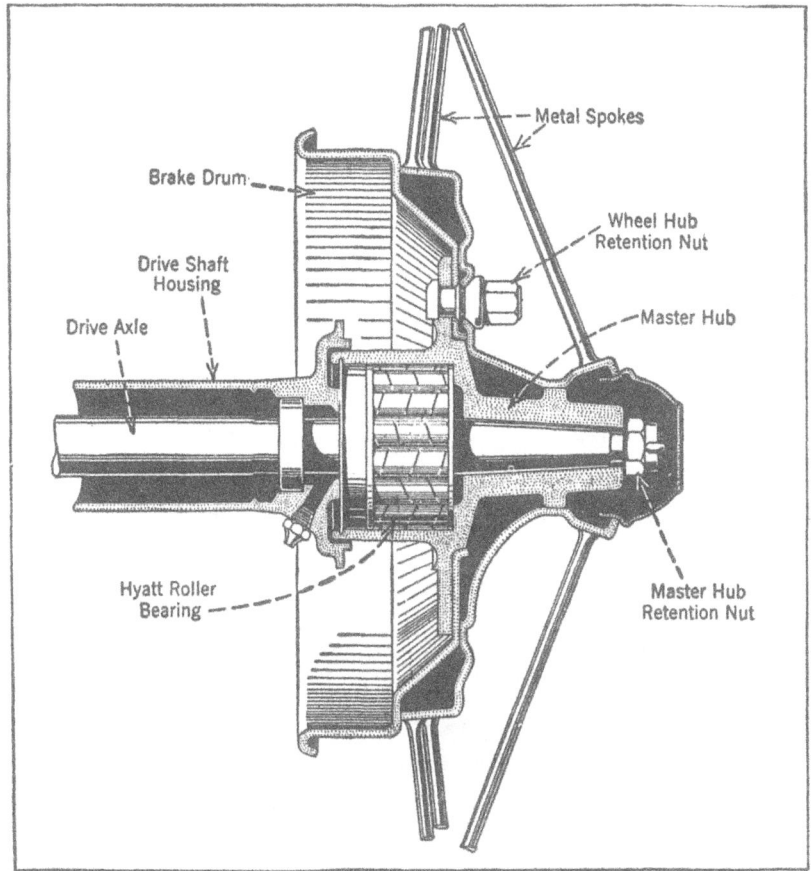

Fig. 13.—Sectional View of Ford Model A Rear Wheel Showing Large Hyatt Roller Bearing for Hub Support.

brake working parts are cadmium-plated, thus making them rust-proof. The brake actuating rods are clearly indicated in diagrams at Fig. 14, which shows the front

The Ford Model A Car

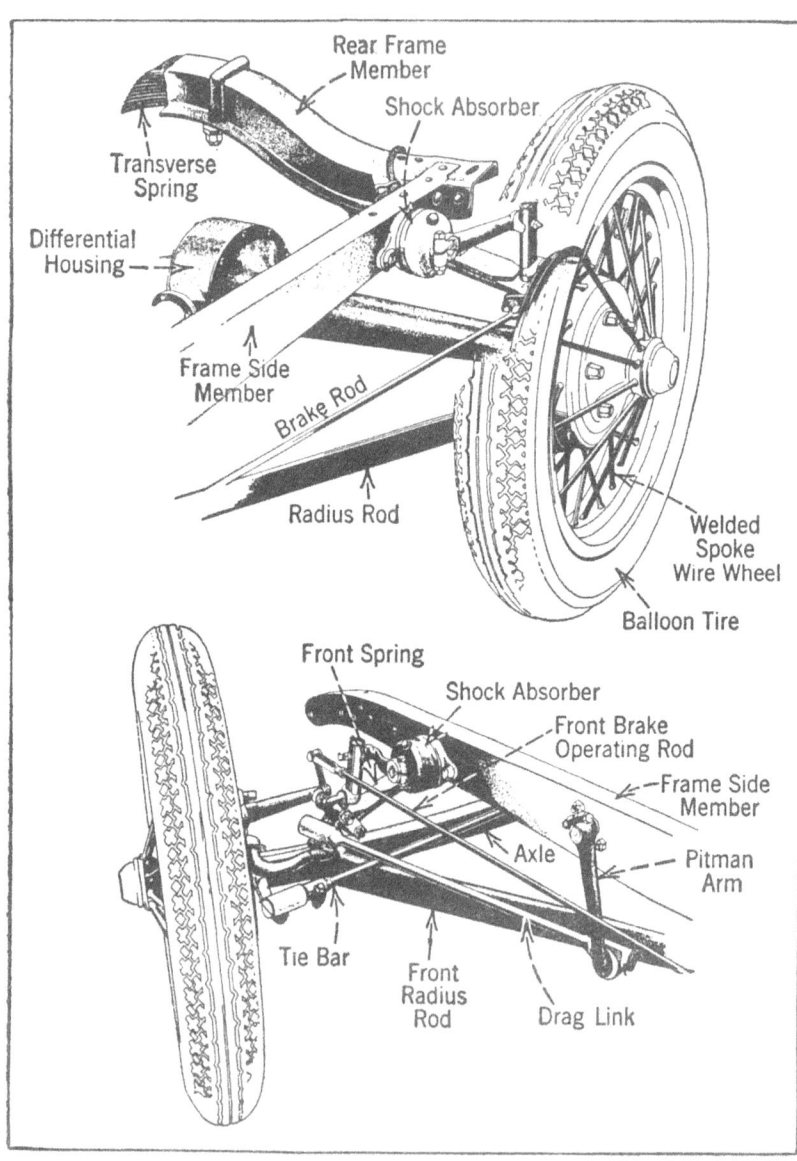

Fig. 14.—View of Rear Corner of Ford Model A Frame at Top Shows Transverse Rear Spring and Shock Absorber Installation. Front Corner of Ford Chassis Showing Steering Members and Front Wheel Brake Operating Rod. Note Hydraulic Shock Absorber.

Details of Model A Chassis

and rear spring construction and the installation of the shock absorbers.

As formerly, the springs are of the transverse semi-elliptic type. They are built of fine spring steel and the leaves are unusually wide and very thin. Each spring is built up of varying sizes and numbers of leaves to give proper flexibility for the particular body style for which the chassis is designed. The action of the hydraulic shock absorbers is adjustable and controls both up and down motion, thus resulting not only in greater comfort to passengers, but also making the car safer. Torque tube drive, which takes up all strain of starting and stopping the car, leaves the rear springs free to perform their sole function—that of carrying the car and passengers. The lubrication of the chassis is by means of a grease gun and special fittings.

Two-shoe construction is used for the internal four-wheel brakes. These "Grey Rock" shoes carry the regulation riveted woven brake-lining. The brakes are rod-operated. The brakeshoes are spread apart by a wedge between rollers on the shoes, as shown in Fig. 15, the wedge push rod passing through the knuckle pin. At the top the shoes are connected by short levers to a bracket which has an adjustment outside of the brake, this adjustment spreading the tops of the shoes apart to take up wear. The brake operating pressure is equalized between front and rear. From the pedal a short rod runs back to a cross-shaft under the center cross-member, the pull being transmitted by means of bell cranks to two horizontal shafts below them. The outer ends of the latter shafts are not rigidly supported, but can move back and forth. Attached to these outer ends are double-

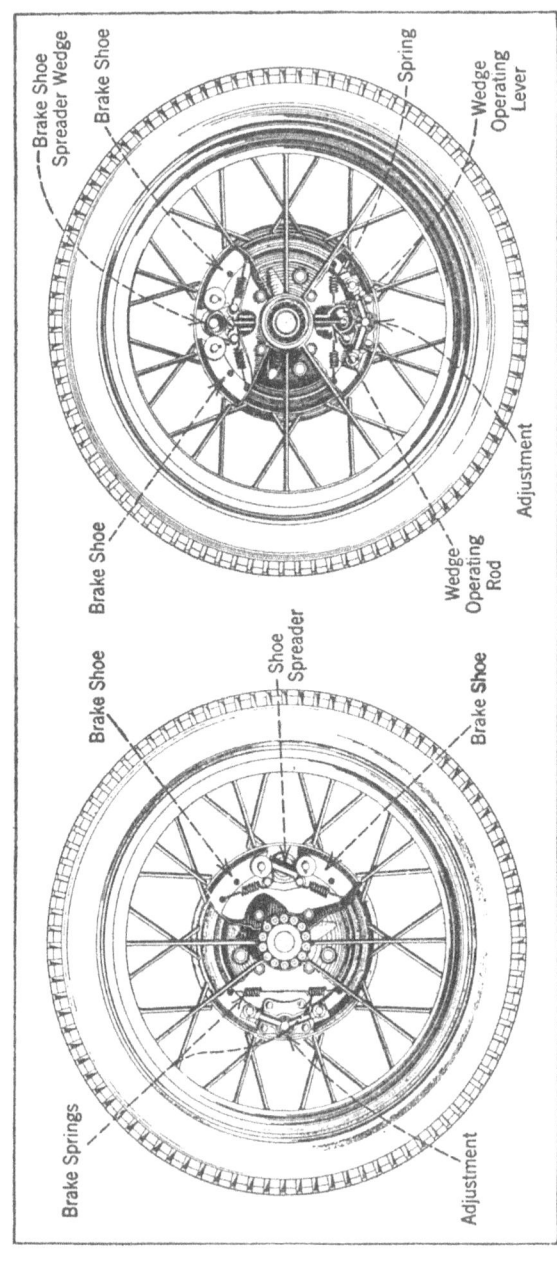

Fig. 15.—View of Rear Wheel of Ford Model A Car at Left Shows Expanding Two-Shoe Brake. The Illustration at the Right Shows the Method of Actuating Front Wheel Brake by a Rod Passing Through Steering Spindle Bolt.

Details of Model A Truck

armed levers, the upper arm actuating the rear wheel brake.

The car has a wheelbase of 103½ in. and a tread of 56 in. The rear axle reduction is 3.7 to 1. The minimum road clearance is 9½ in. The center section of the front axle is made of chromium steel. Brake diameters are 11 in. all around and the length of the lining is 14 in. per wheel, the width of the lining being 1½ in., and the thickness 3/16 in. The total braking area is 168 sq. in. The front spring is 1¾ in. wide; the rear, 2¼. All springs are made of chromium steel. The pressed steel frame is of 5/32 in. stock, and the section is 4 in. deep.

The Ford Model A Truck.—Truck models in the new Ford line are rated at 1½ tons with single rear wheels, and 2½ tons with dual rear wheels. Four-wheel brakes, worm-drive rear axles, torque tube drive, cantilever rear springs, and welded steel-spoke wheels are used on both models, which are practically identical except for rear wheel equipment. The powerplant is the same as that embodied in the new passenger cars consisting of a four-cylinder 3⅞ by 4¼ in. engine with battery-distributor ignition and oil and water pumps, multiple-disk clutch and three-speed sliding gear transmission. An additional two-speed auxiliary transmission is offered at extra cost. Front axle, front wheel brakes and steering gear are substantially the same as the passenger car units.

An inclosed cab and express, stake and platform bodies are available. Cabs are all steel and are finished in imitation leather, and doors, which are equipped with crank window regulators, are similar to those used on the passenger cars. The gearshift lever is in the center and the hand brake lever on the left. The light control switch is placed above the steering wheel and the spark

The Ford Model A Car

and throttle levers are below it, but quadrants are not provided. A carburetor control rod is beneath and forward of the cowl on the right, and the gasoline shut-off valve may be reached from the driving compartment. The express body is all steel except the floor and sills, which are of seasoned wood. This body is 7 ft. 2 in. long and 4 ft. wide, while the stake body measures 8 ft. 1½ in. inside length and 5 ft. 8 in. width, the racks extending 26 in. above the floor.

Distance from the back of the cab to the rear of the frame is approximately 81¼ in., and from back of cab to center of rear axle about 52 in. The height of the frame is approximately 31 in. at the rear and 27½ in. at the rear of the cab. The frame tapers outward from the rear, being about 33½ in. wide at the cab and 40½ in. at the rear end. The rear axle is of the three-quarter floating type, with rear wheel loads taken by roller bearings on the outside of the housing. The assembly embodies two axle housing halves bolted together and an overmounted worm with double external adjustments. Standard gear ratio is 5 to 1. Two radius rods are employed on the torque tube and the axle end of each rod is used to support the inner end of the brake shaft. Two rear wheel brake drums, which are 14 in. in diameter, inclose internal brakes of the rigid-shoe type. Two shoes are used in each brake, and they are divided on a horizontal line with the operating cam at the front and the adjustable anchorage at the rear.

The frame has an overall length of 171⅝ in., and the depth is 6 in. with flange width of 2¾ in. The bottom of the side rail tapers upward from the trunnion mountings for the rear springs to the rear end. Five heavy channel cross-members are used, the rear member being

Model A Truck Chassis Details

offset downward and fitted with a bracket for carrying the spare wheel, or wheels, below the body. The next member forward is of U-section with wide flanges, and is arched to provide clearance for the axle housing. Third carries the forward end of the torque tube and the next carries the front propeller shaft, or the bearing between the two transmissions if the dual unit is used. The front cross-member mounts the engine front support, the radiator and the transverse front spring. In addition, the rear engine supports virtually form an additional cross-member.

Wheel base is $131\frac{1}{2}$ in., compared with 124 in. on the former truck. The rear axle is similar to the unit formerly used, and the necessary extension of the power line is accomplished by use of an inclosed intermediate propeller shaft. At the rear of the transmission a tube with a ball end similar to that on the front of the torque tube is mounted within the ball cap, and this tube and the propeller shaft extend to a frame cross-member where the universal joint, ball and cap assembly are duplicated. The auxiliary transmission, when supplied, is placed back of the regular unit and in place of the front part of the intermediate propeller shaft.

The 17-leaf rear springs are located outside the frame rails and are shackled to the frame at the front end and pass below the trunnion mounting on the side of the frame. Connection with the rear axle housing is made by a split member which encircles the housing. The main leaf of the spring is practically level and has little if any camber.

Auxiliary Two-Speed Transmission.—The dual transmission is of the planetary constant-mesh type and provides a reduction of 1.47 to 1. It has three planet pinions

The Ford Model A Car

which turn around a sun gear and mesh with the large internal gear; being similar in this respect to the Model T steering gear. This auxiliary transmission is controlled by means of a separate internal-and-external gear clutch provided for this purpose ahead of the planetary gear-

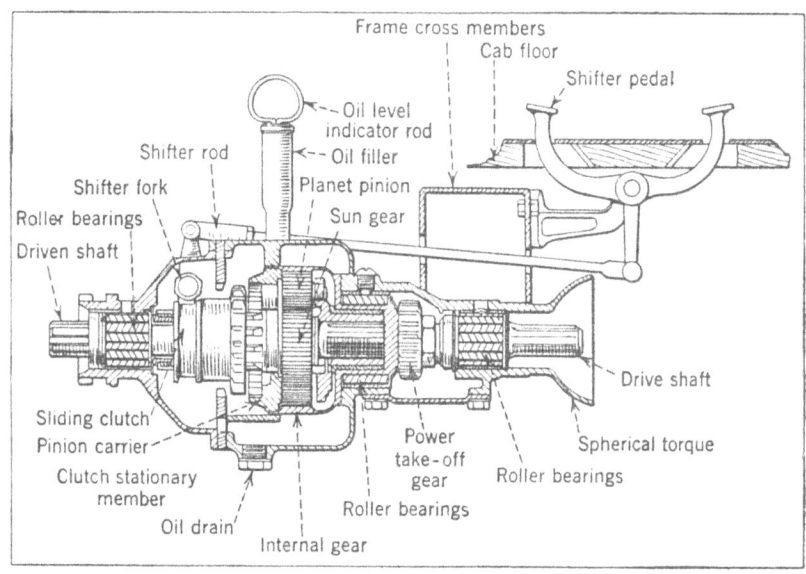

Fig. 16.—Longitudinal Section of the Auxiliary Two Speed Transmission for Ford Model A Truck Showing Planetary Gears and Positive Clutches.

ing. In direct drive, of course, there is no relative motion of the working gears, since the sliding member of the clutch mentioned locks the planetary system, which latter is mounted on roller bearings as shown by the accompanying illustration, Fig. 16.

For low speed, the clutch member locks the sun gear of the planetary system so it cannot revolve. The drive is then taken from the engine through the large internal

Auxiliary Two-Speed Transmission

gear of the planetary system, forcing the three planet pinions to revolve around the sun gear. Since the sun gear is stationary, the planets do not rotate around the axis of the transmission as rapidly as the internal gear, thereby providing a reduction which in this transmission is about 32 per cent. The cage which supports the three

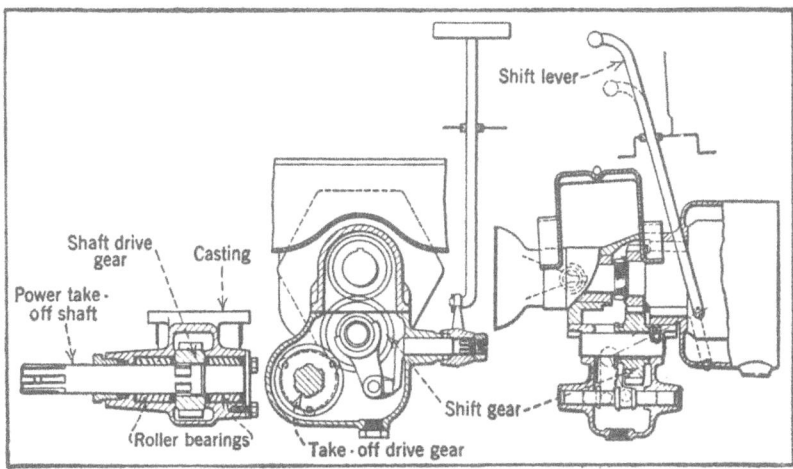

Fig. 17.—Suggestion for Power Take-Off Arrangement to Use in Connection with Ford Auxiliary Transmission.

pinions is connected to the driven shaft, which transmits the torque to the rear axle.

The auxiliary transmission is foot-controlled. There are two pedals projecting through the cab floor. By depressing one pedal, the auxiliary transmission is locked and therefore disengaged, while by depressing the other pedal, it is engaged. It is claimed that the pulling capacity of the truck is increased 47 per cent by the use of this transmission. In direct drive, speed and pull are the same as with the standard truck without the auxiliary transmission.

The Ford Model A Car

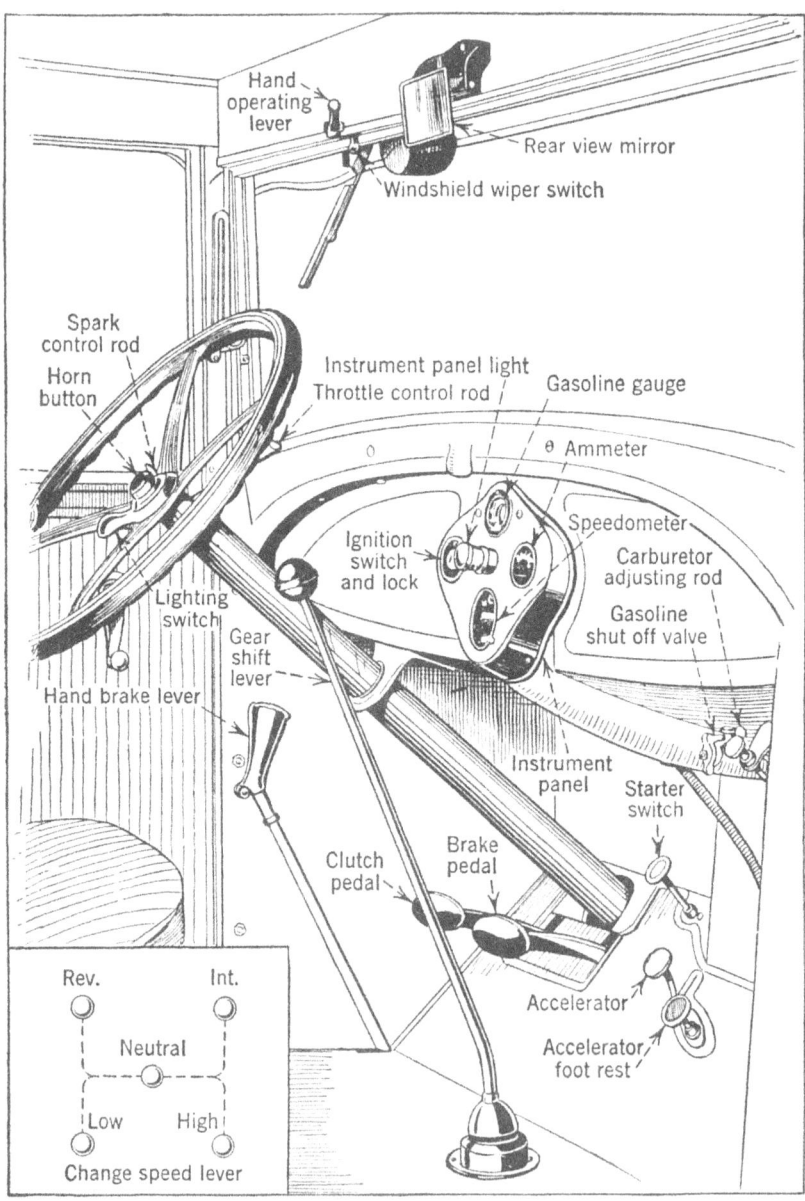

Fig. 13.—Instruments and Control Levers of Ford Model A Cars Follow Conventional Practice. Inset Shows Standard Gear Shift Lever Positions.

Operating Ford Model A Car

This transmission is provided for the new Ford truck as optional equipment at extra cost. It is also provided with a power take-off gear for use with dump bodies, etc. The power take-off itself is not supplied by the Ford Motor Co., which furnishes the suggested design reproduced herewith at Fig. 17.

OPERATING FORD MODEL A CAR

Filling the Radiator.—Before starting the car, see that the radiator is filled with clean fresh water. The cooling system holds approximately three gallons. If in winter, use anti-freeze solution (procurable from any Ford dealer). As the proper cooling of the engine is dependent upon the water supply, it is important particularly with a new car to see that the radiator is kept well filled. Any steaming causes loss of water as will leaks in hose or radiator.

Filling the Gasoline Tank.—The tank has a capacity of ten gallons. The gasoline gauge on the instrument board panel indicates the amount of gasoline in the tank. Gasoline can be drained from the tank by opening the petcock in the sediment bulb located on the engine side of the dash. The screen in the gasoline tank filler neck should occasionally be removed and cleaned. IMPORTANT— After cleaning be sure to replace this screen. Also see that the small vent hole in the gasoline tank cap does not become clogged. If the vent hole is clogged the gasoline will become air bound and fuel supply will be irregular.

Proper Oil Level. — Before starting the engine, make sure there is a sufficient supply of medium light, high grade engine oil in the oil pan. If there is not enough oil, more should be added through the breather pipe lo-

The Ford Model A Car

cated at the left side of the engine (a metal cap covers it). Five quarts of oil is the amount required in the oil pan. To determine the correct oil level, use the indicator, which is of the bayonet type, located on the left side of the engine just to the rear of the breather pipe as follows: Pull out the indicator—wipe it off—re-insert the indicator and again remove it. The mark made by the oil indicates its level. When the oil reaches the point marked "F" on the indicator, it is at the proper level. Under no circumstances should the oil be permitted to get below the point marked "L" as any attempt to run the engine with too little oil may seriously damage the parts. The letter "F" indicates "full," the letter "L" means "low." When replacing the oil level indicator, see that both the short and long ends of the indicator enter the opening in the crankcase. Failure to insert both ends into the opening permits the oil to leak out as the opening is not fully closed.

Before Starting the Engine. — Be sure the gear shift lever is in neutral position—i.e., the position in which it can be moved freely from side to side, as in any standard gear shift car.

Advance the throttle lever, located under the steering wheel (right hand side), about three notches, or until the accelerator pedal moves slightly downward. The Ford Model A control levers and instruments are shown at Fig. 18.

Pulling down the throttle lever or pressing on the accelerator pedal, controls the quantity of gas entering the cylinders, and regulates the speed of the engine. In the Model T there was no accelerator pedal. Carburetor control was entirely by hand lever.

Place the spark lever (left hand) at the top of the

Operating Ford Model A Car

quadrant (the notched quarter-circle on which the lever is operated). This is the retard position. The spark lever regulates the timing of the spark which explodes the gas in the cylinders. Readers who are familiar with the Model T will note the changed location of engine controls.

Always retard the spark lever when starting your car. Starting the engine with the spark advanced may cause the engine to kick back, and damage the starter parts. After the engine is started, advance the spark lever all the way down the quadrant.

Theft-Proof Lock.—The Ford type electro-lock used in the ignition switch is a combination switch and theft-proof lock affording full protection for the car and meeting the exacting requirements of the underwriters as regards theft insurance. To unlock the electro-lock, simply insert the switch key into the ignition switch and turn the key to the right. This releases the cylinder of the lock, which snaps forward and closes the ignition circuit. When the cylinder is released the engine can be started in the usual manner, AND THE SWITCH KEY WITHDRAWN FROM THE LOCK.

To shut off the engine, push in on the cylinder of the lock until it snaps back in the lock position. This shuts off the ignition and locks the car in a positive manner.

Starting the Engine.—1. Release the lock cylinder by turning the switch key to the right. 2. See that the spark lever is retarded; the throttle lever advanced three or four notches on the quadrant, and the gear shift lever in neutral position. 3. If the engine is cold, turn the carburetor adjusting rod one full turn to the left, to give it a richer mixture for starting. This rod serves both as a choke for starting and as an enriching adjustment.

The Ford Model A Car

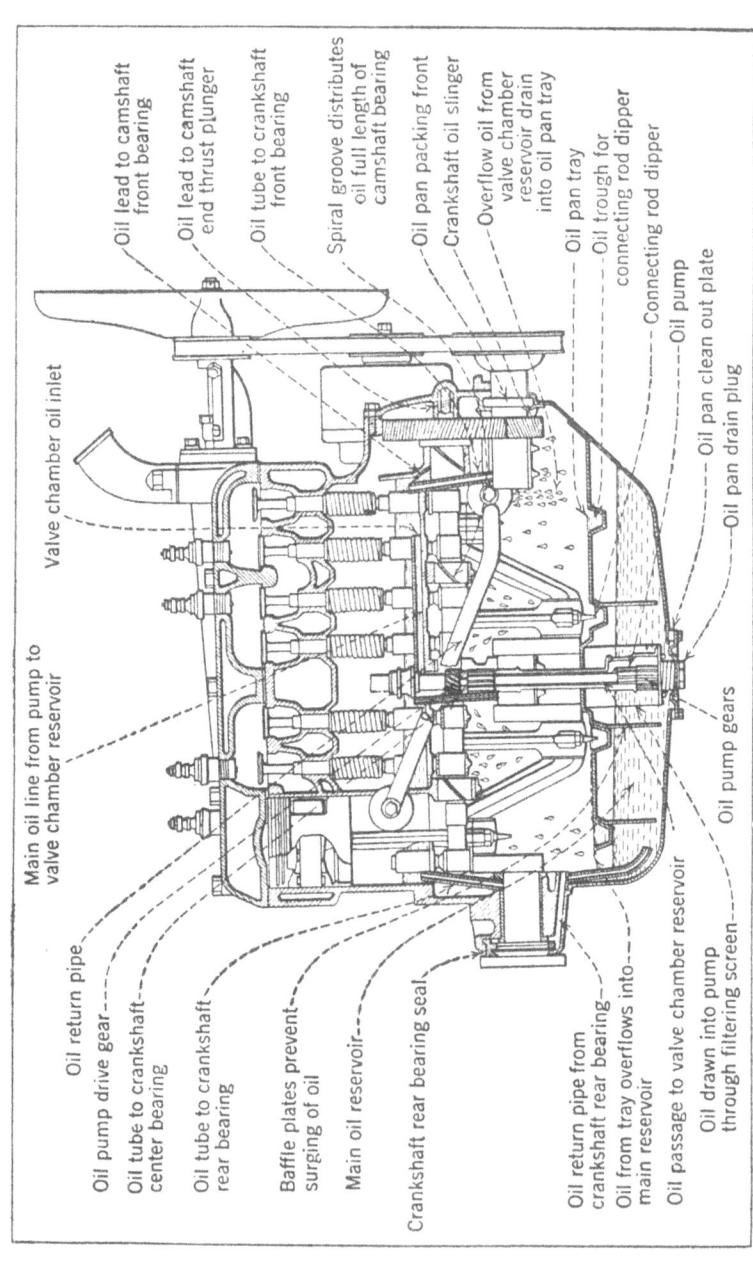

Fig. 19.—Sectional View of Ford Model A Engine Showing Method of Lubricating Internal Mechanism by Combined Pump and Splash System.

Starting the Model A Car

Next pull back the rod, at the same time pressing down on the starter button and release the choke rod; next advance the spark lever. When the engine warms up, turn the choke rod back to its normal position one-quarter turn open.

When starting a warm engine, do not pull back the choke rod unless the engine fails to start on the normal mixture, as there is a possibility of flooding the engine with an over-rich mixture of gas. If you should by accident flood the engine, open the throttle and, with the choke rod in normal position, turn the engine over a few times to exhaust the rich gas. If you continue to turn the engine with the choke rod out the flooding will become greater and engine increasingly difficult to start.

Starting the Car.—Release the hand brake lever. With the engine running, disengage the clutch by pushing down the left foot pedal. Move the gear shift lever to the left and back, which is the low speed position as shown in insert in Fig. 18. Gradually release pressure on the clutch pedal, allowing it to return to its normal position, and at the same time increase the speed of the engine by pressing lightly on the accelerator. Second Speed: When the car has reached a speed of five to eight miles per hour, engage second or intermediate speed as follows: Release pressure on accelerator and again disengage the clutch, moving the gear shift lever through neutral to the right and forward, second speed position. Allow the clutch pedal gradually to return to its normal position, and increase the speed of the car until it is running 12 to 15 miles per hour. High Speed: Disengage the clutch as before; at the same time release the pressure on the accelerator and pull the gear shift lever straight back from second speed. Then engage the clutch and increase

The Ford Model A Car

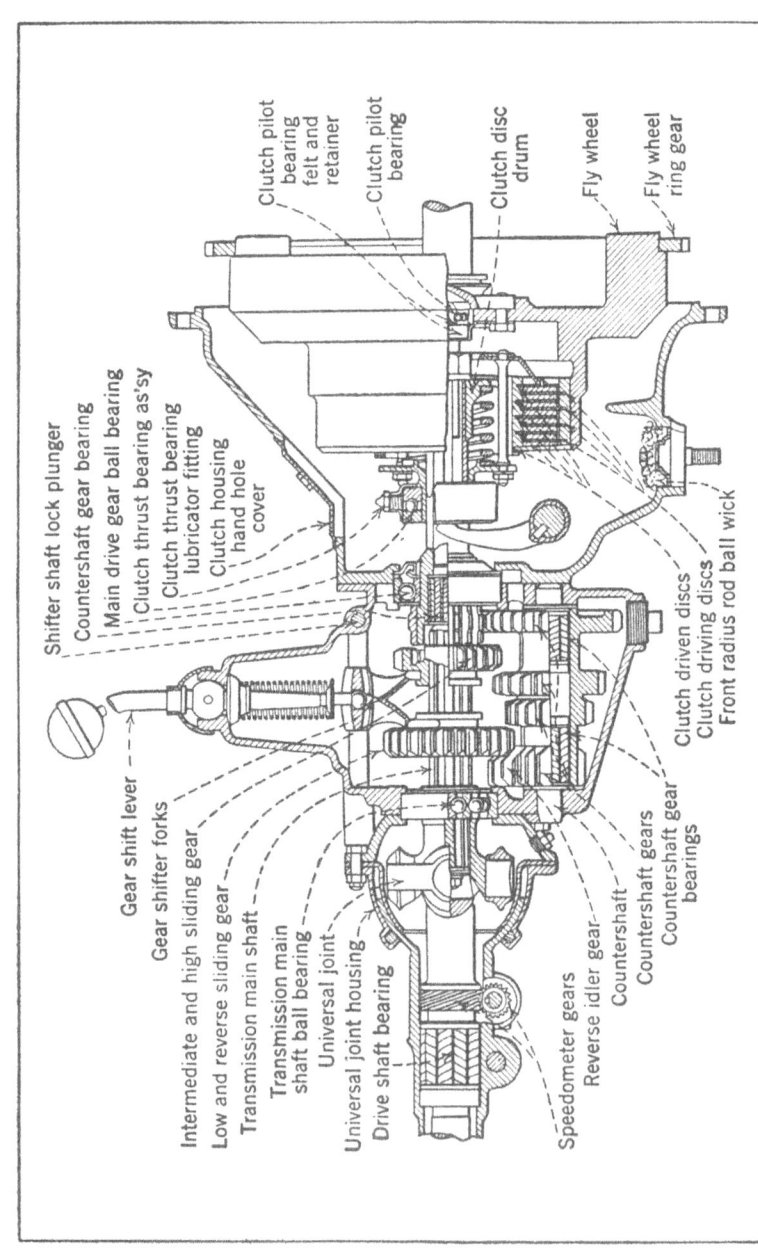

Fig. 20.—Sectional View of Clutch and Gearset Assembly of Ford Model A Showing Clutch and Change Speed Gear Relation.

Ford Model Clutch and Gears

the speed of the engine, as driving conditions may require.

Shifting Back Into Low Speed.—When shifting from high to second speed, at car speeds not exceeding 15 miles per hour, there should be no hesitation in neutral. The lever should be moved as quickly as possible, from high to second speed. Should it be necessary to shift from high to second speed at higher car speeds, it can be done by the following method: Disengage the clutch and shift into neutral. Reengage the clutch and at the same time accelerate the engine; then disengage the clutch again and shift to second, after which disengage the clutch. With a little experience, this shift can be made quietly.

Descending a Hill. — When descending long, steep grades, have the transmission in gear, the clutch engaged, and the ignition switch on. This allows the engine to turn over against compression and act as a brake. On very steep grades the car should be in second speed gear before descent is started. On exceptionally steep grades the low speed should be used. This increases the braking action of the engine. Always leave the ignition switch on when descending an incline. Shutting off the switch allows raw gas to be drawn into the cylinders, which washes the lubrication off the cylinder walls. Also unexploded gas collects in the muffler, and when the switch is again turned on there is a possibility of blowing out the muffler, which has become filled with highly explosive mixture.

To Stop the Car.—Disengage the clutch by pressing forward on the left pedal and apply the foot brake by pressing forward on the right pedal. Except when a quick stop is necessary, it is advisable to apply the brake gradually. When driving on wet or slippery pavement the speed of the car should be reduced by applying the foot brake before releasing the clutch. This method of braking prolongs the life of the brake lining, and is a safety factor.

In bringing the car to a final stop, keep the clutch disengaged until the gear shift lever has been moved into neutral position. To stop the engine, push in on the electro-lock cylinder until it snaps back into the lock position. The driver should endeavor to so familiarize himself with the operation of the car that to disengage the clutch and apply the brakes becomes automatic—the natural thing to do in case of emergency.

To Reverse the Car.—The car must be brought to a stop before attempting to reverse its direction. To shift into reverse, proceed as in shifting into low speed, except that the gear shift lever is moved to the left and forward.

The Spark Control.—For normal driving, drive with the spark lever fully advanced. When the engine is under a heavy load, as in climbing steep hills, driving through heavy sand, etc., the spark lever should be retarded sufficiently to prevent a spark knock. This is a pronounced knock, which is sometimes mistaken for a worn bearing.

Driving the Car.—The different speeds as required to meet road conditions are obtained by varying the pressure on the accelerator, which in turn controls the motor. Practically all the running speeds needed for ordinary travel are obtained in high gear; the low and second speed gears are used principally to give the car momentum in starting, and when the engine is subjected to a heavy load.

Caring For the New Car.—A new machine requires

Driving Ford Model A Car

more careful attention during the first few days it is being driven than after the parts have been thoroughly "worked in." To obtain the best results, a new car should not be driven faster than 30 to 35 miles per hour for the first 500 miles. The oil in the engine should be changed after this distance has been covered, as there may be a few metallic particles or sand in a new engine. Never start out with your car until you are sure it has plenty of oil, water and gasoline.

See that an air pressure of 35 pounds is maintained in all tires. Under inflation causes more tire expense than anything else.

Inspect your battery every two weeks and keep it filled to the proper level with distilled water. If the water is allowed to evaporate below the top of the plates the life of the battery will be seriously affected.

Let the Ford dealer or expert Ford repairman go over your car once a month, making any mechanical adjustments necessary to keep your car in proper running order.

Do Not Rest Foot on Clutch Pedal. — Do not make a practice of resting your foot on the clutch pedal while driving, as this may cause the clutch to slip and unnecessarily wear the facing on the discs. This is called "riding the clutch" and causes premature clutch trouble in any make of car. The correct clearance or play for the clutch pedal is approximately ¾ in. That is when the clutch pedal is depressed there should be about ¾ in. play in the pedal before it starts to disengage the clutch. As the clutch facings wear this clearance or play gradually grows less. Consequently it should occasionally be checked. Under no circumstances should the car be driven without clearance or play in the clutch pedal.

The Ford Model A Car

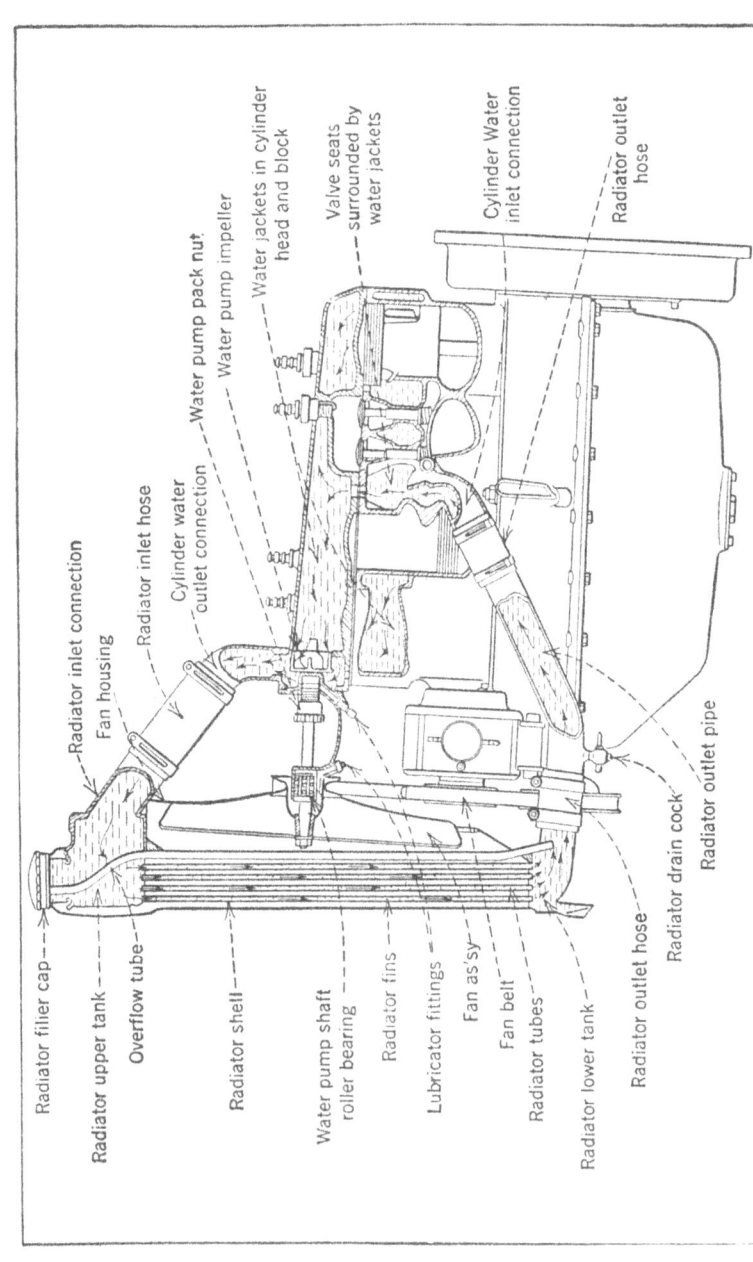

Fig. 21.—The Ford Model A Engine Water-Cooling System Illustrated in Detail.

Cooling Fan Belt Adjustment

The adjustment is easily made by removing a steel clevis pin and turning the release arm rod after it is lifted

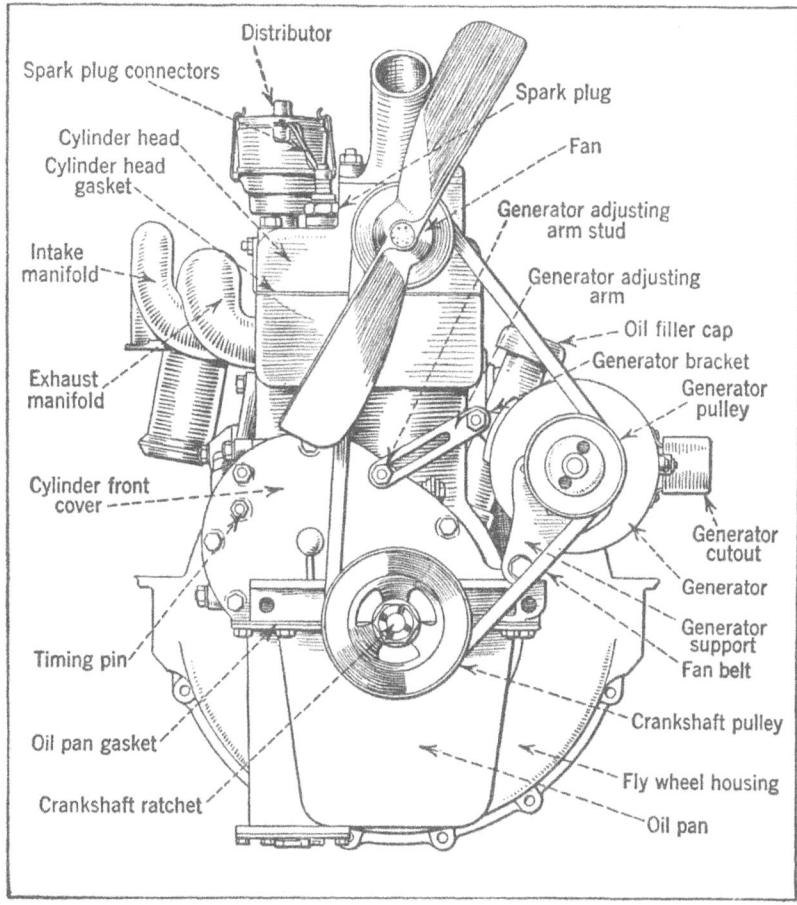

Fig. 22.—Front View of Ford Model A Engine Showing Cooling Fan and Generator Drive by V Belt, Also Method of Adjusting Belt Tension.

from the actuating lever. Screwing the rod in augments the clutch pedal play. Screwing the rod out reduces the play. After making adjustment, be sure to replace steel pin and cotter key.

The Ford Model A Car

The Ford Lubricating System.—The purpose of lubricating is to reduce friction between moving surfaces. The oiling chart gives full information for lubricating the Ford car and can be obtained from any agent. Proper lubricating has a vital effect on the life of your car; consequently, you should follow the instructions of the manufacturer very carefully.

All parts of the engine are lubricated from the oil reservoir in the oil pan by the Ford Pump, Splash and Gravity Feed. Only medium light, high grade engine oil should be used in the engine.

Oil of the proper kind reaches the bearing surfaces with greater ease and cuts down friction and reduces heating. It should, of course, have sufficient body so that the pressure between the two bearing surfaces will not force out the oil and allow the metal to come in actual contact. Too light in body is as undesirable as too heavy body in oil. Heavy and inferior oils have a tendency to carbonize quickly, also "gum up" the piston rings, valve stems and bearings. In cold weather a light grade of oil having a low cold test is absolutely essential for the proper lubrication of the car. Such oils can be secured from any Ford service station or reliable supply house.

Oil conforming to the following specifications is recommended for use in the Model "A" engine. The engine lubricating system is shown at Figure 19.

Draining the Oil Pan.—It is advisable to clean out the oil pan by draining off the dirty oil when the new car has been driven 500 miles, and thereafter to repeat this operation every 500 miles. The oil should be warm before draining. It is a good plan, especially in cold weather, to remove and clean the oil pan clean-out plate once a month.

Model A Lubrication System

WINTER OIL

COLOR	FLASH	FIRE	VISCOSITY		COLD SET	GRAVITY
Not darker than	(Open Cup) °F.	°F.	Saybolt-Universal 100 °F.	210 °F.	°F.	Beaume
No. 5	390 Min.	450 Min.	450 Sec. Max.	58 Sec. Max.	20 Max.	22 Min.

SUMMER OIL

COLOR	FLASH	FIRE	VISCOSITY		COLOR SET
Not darker than	(Open Cup) °F.	°F.	Saybolt-Universal 100 °F.	210 °F.	°F.
No. 6	400 Min.	450 Min.	650 Max.	66 Max.	45 Max.

Lubricating the Differential.—Every 5,000 miles the lubricant in the differential should be drained and the housing flushed with kerosene. New lubricant should then be added until it reaches the level of the oil hole in the housing

Lubricating the Transmission.—About once every 5,000 miles the gear lubricant should be drained from the transmission by removing the drain plug at bottom of transmission case. The interior of the transmission case should be then thoroughly flushed with kerosene and refilled with fresh gear lubricant. The new lubricant is poured into the transmission through the filler hole, located at the right hand side of the transmission case. Pour sufficient lubricant in until it reaches the level of the filler hole.

Lubrication of the Clutch Bearing. — The clutch pilot bearing at the front end of the clutch is thoroughly packed with grease when the car is assembled, and it will not be necessary to lubricate this bearing until such time as the clutch may be disassembled. When the

The Ford Model A Car

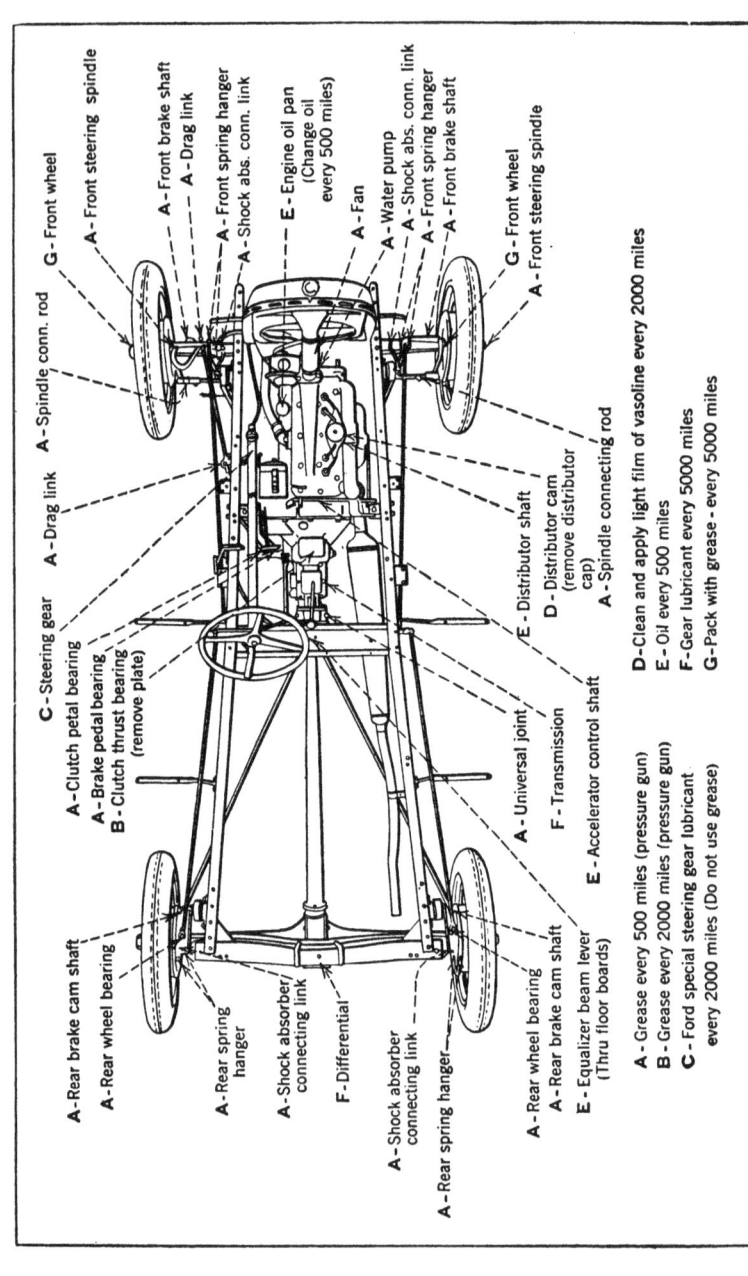

Fig. 23.—Diagram of Ford Model A Chassis Showing Parts That Need Periodical Lubrication.

Greasing the Car

clutch is disassembled the bearing should be repacked with a good grade of cup grease.

Approximately every 2,000 miles, lubricate the clutch release bearing. This is done by removing the hand hole cover (see Fig. 20) and turning the bearing until the lubricator fitting is at the top. Lubricate the bearing by means of the compressor grease gun.

IMPORTANT NOTE.—The clutch is a dry disc clutch and under no circumstances should it be oiled.

Greasing the Car.—In order to properly force lubricant to all parts equipped with the conical-shaped fittings, a high pressure lubricating system is employed. With this system the lubricant can be forced in under a pressure of 2,000 pounds or more per square inch, thus assuring a more thorough and positive lubrication than can be accomplished any other way. A compressor is supplied with the tool equipment of each car, and by means of this device lubricant can be forced into all bearings provided with conical-shaped fittings.

Remove top cap and plunger assembly to fill the compressor. Fill the barrel with lubricant. Pack the lubricant solidly. To avoid air pockets, tap the nozzle gently on a board or work bench while filling. To prevent lubrication backing up and soiling hands, FILL UP ONLY TO THE TOP OF THE LETTERING ON THE OUTSIDE OF THE BARREL.

When the compressor is pressed against the conical-shaped fittings, the plunger moves forward, forcing the lubricant in the nozzle directly through the fitting into the bearing, under an extremely high pressure. When the pressure on the handle is released, grasp the barrel of the compressor with one hand and draw back the handle with the other, so as to load the compressor and

The Ford Model A Car

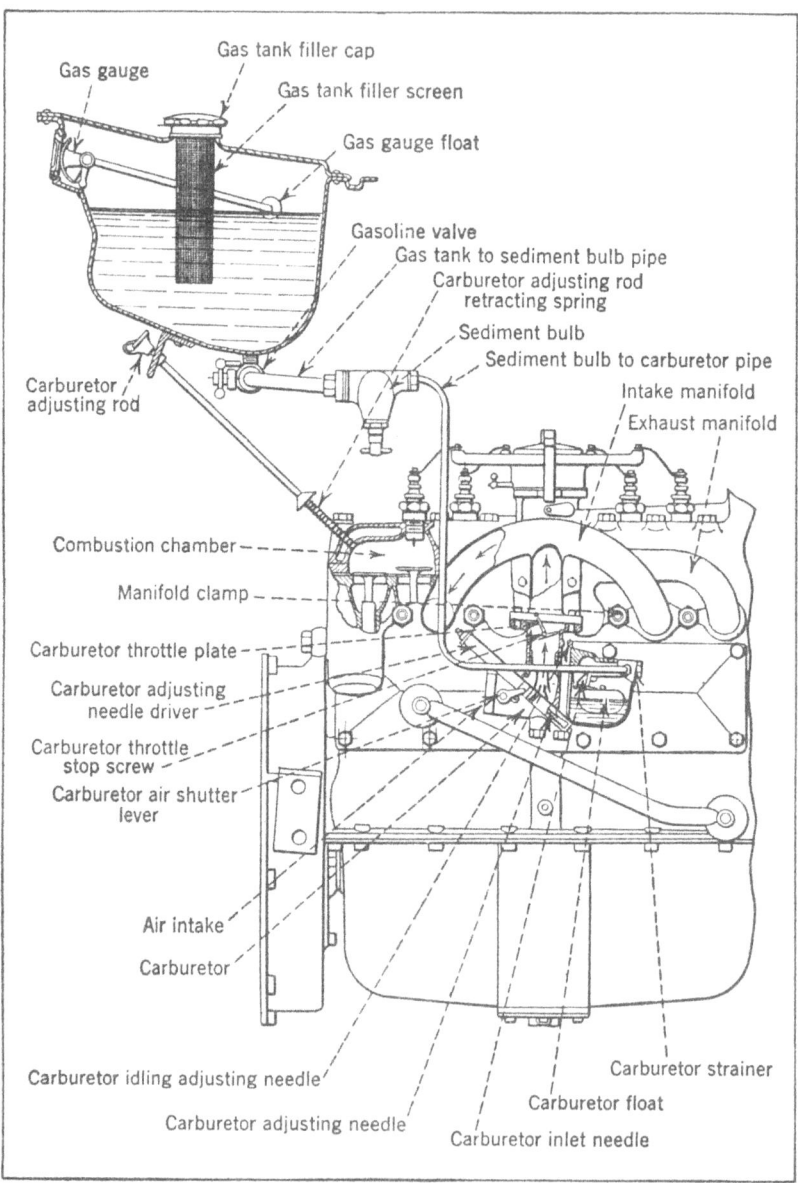

Fig. 24—The Ford Model A Engine in Part Section Showing Units That are Part of the Fuel Supply and Carburetor System.

Model A Cooling System

make it ready to deliver a charge of lubricant with the next forward thrust.

The bearings in the generator and starting motor are lubricated when they are installed in the car, and require no further attention. The distributor should be kept clean and well oiled. Put oil in the oil cup at the side of the distributor every 500 miles. Add sufficient oil to reach the level of the oil cup. Every 2,000 miles remove the distributor cap, clean the lobes of the cam and apply a light film of vaseline.

Cooling the Engine.—The Ford engine is cooled by a circulation of water through the cylinder water jackets which surround the combustion chamber and valve seats. The water is circulated by thermo-syphon action, the flow of water being accelerated by means of a centrifugal water pump located in the front of the cylinder head. This pump draws the heated water from the engine into the upper radiator tank, where it is cooled by filtering through the radiator tubes to the lower tank. The radiator is cooled by means of the fan located just back of the radiator, where it draws a current of air around the radiator tubes. To prevent overheating, keep the radiator well filled. The capacity of the radiator is three gallons and the cooling system parts are clearly shown in illustration at Fig. 21.

Adjusting the Fan Belt.—The fan and water pump both operate from the same shaft. The shaft is driven by a "V" shaped rubber belt, as shown in front view of engine at Fig. 22. The belt is adjusted to the proper tension when the car leaves the factory, and this adjustment should not be changed unless the belt slips. The adjustment is easily made by loosening the generator adjusting arm stud nut, and moving the generator out-

The Ford Model A Car

ward. Do not tighten the belt more than is actually necessary to keep it from slipping, as too much pressure from excessive tension will cause undue loading on the pump shaft.

Packing is used in forming a water-tight connection around the water pump shaft. Should the packing become loose, tighten the packing nut. A screw driver is used for this purpose, using it as a pry and resting the blade on the supporting casting. Do not tighten the nut more than is necessary to stop the leak. The point of the screw driver is inserted in one of the slots of the stuffing box nut and rocked about the fulcrum point by bearing on the lever.

The entire circulating system should be thoroughly flushed out several times each season. To do this, open the petcock at the bottom of the radiator outlet connection pipe and insert a hose into the filler neck, allowing the water to flow through the system for about fifteen minutes, or until the water comes out clear.

Care of the Radiator in Winter.—In freezing weather it is necessary to use an anti-freeze solution in the circulating system to prevent freezing of the water and bursting the tubes in the radiator. This cooling solution should be put in with the first approach of cold weather.

Do not overlook the fact that constant evaporation will eventually weaken most anti-freeze solutions, consequently they should be tested frequently, especially in severe weather. It is better to have too rich a mixture of solution than to have too little compound. A suitable anti-freeze solution can be obtained from any authorized Ford dealer. Also complete directions as to the percentage of solution to be used to withstand the varying degrees of cold. As anti-freeze solutions usually contain

Model A Fuel System

alcohol, care should be used when filling the radiator not to spill any of the solution, as it may damage the pyroxylin lacquer finish used on Ford cars. After filling the radiator, be sure the radiator cap is screwed down tight.

Fuel System.—The gasoline is carried in a ten-gallon tank welded integral with the cowl of the car. From this tank the gasoline flows by gravity to the carburetor. There it is mixed with air and drawn into the cylinders through the induction manifold and intake valves by piston suction. A sediment bulb located on the engine side of the dash is provided for draining off water or sediment that may have accumulated in the tank. Occasionally draining the bulb prevents foreign material being drawn into the carburetor.

The Carburetor.—The quantity of gasoline entering into the carburetor is governed by the needle valve. The volume of gas mixture entering the intake manifold is controlled by opening and closing the throttle, according to the engine speed desired by the driver. Since, with the exception of the needle valve and idle adjustment, all of the carburetor adjustments are fixed, about the only thing that could affect the carburetor would be dirt or water getting into it. An occasional cleaning will insure uninterrupted service. To clean the carburetor, remove the filter screen and thoroughly clean the screen by washing it in gasoline. The screen is easily removed by backing out the filter plug at the front of the float bowl. After cleaning, be sure to replace the screen. It is also a good plan to remove the brass plug at the bottom of the carburetor and drain the carburetor for a few seconds, to make sure any water left from condensation will escape.

For economical driving, careful mixture regulation

The Ford Model A Car

is important. Reduce the quantity of gasoline in the mixture by turning the adjusting rod to the right as far as possible without reducing the speed of the engine. This is particularly true when taking long drives, where conditions permit a fair rate of speed being maintained, and accounts for the excellent gasoline mileage obtained by good drivers. Turning the carburetor adjustment too far to the right results in a "lean" mixture—too far to the left results in a "rich" mixture. A lean mixture has too much air and not enough gasoline. A rich mixture has too much gasoline and not enough air. A rich mixture causes excessive carbon and overheating; likewise, it wastes fuel. The mixture should be kept as lean as possible without sacrificing the power of the engine. Too rich mixture causes black smoke and foul exhaust; too thin mixture causes popping back in carburetor.

The method of regulating the carburetor for ordinary driving conditions is to turn the carburetor adjusting rod to the right until the needle just seats, then turn the rod back one-fourth ($\frac{1}{4}$) of a turn. Use care in adjusting the carburetor, as turning the needle down too tightly will result in its becoming grooved and the seat enlarged. When these parts are damaged, it is difficult to maintain proper adjustment of the carburetor until a new needle is used.

Fully retard the spark lever when setting idling adjustment. Set stop screw on throttle lever so that engine will run sufficiently fast to keep from stalling when it is idling. Turn the idling adjusting screw in or out until the engine hits evenly without "rolling" or "skipping" (the correct setting of the adjusting screw is approximately two turns off its seat). Next back off the stop screw until the desired engine speed is obtained. This

Model A Electrical System

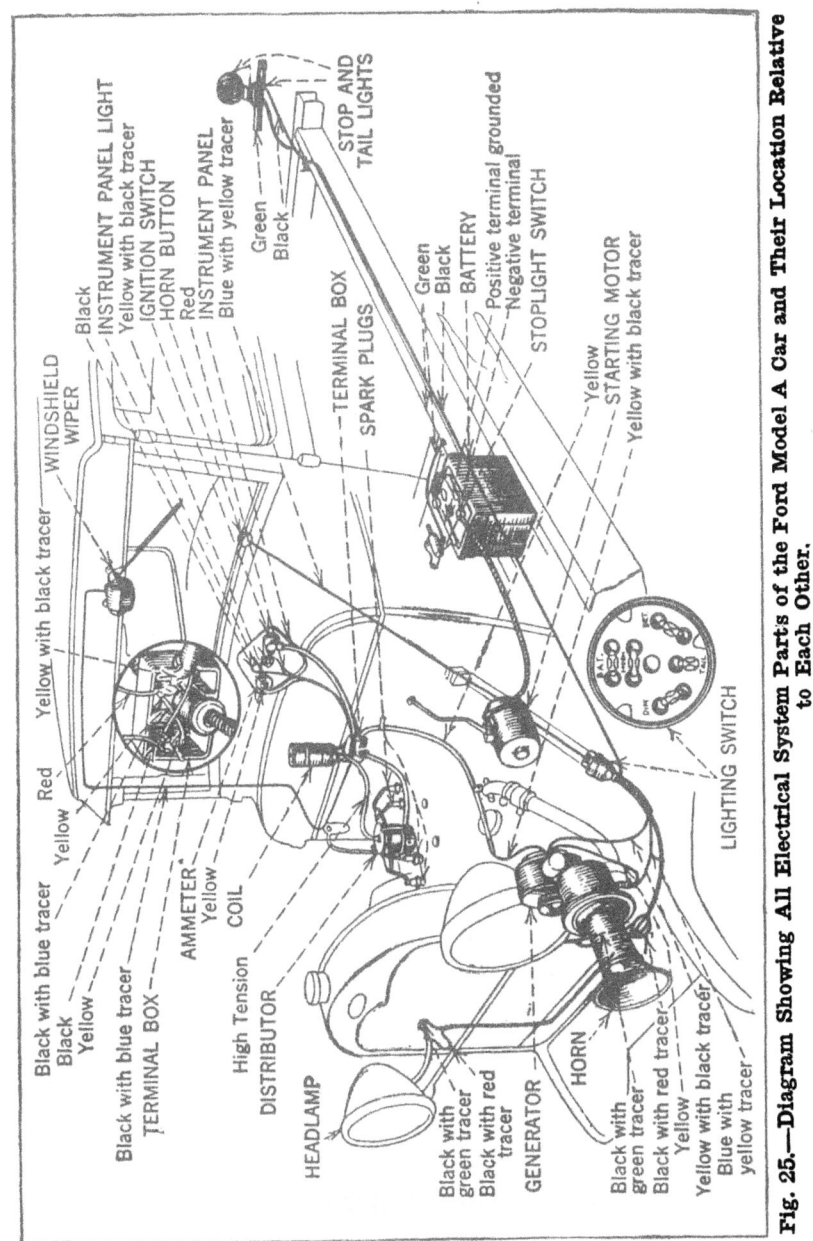

Fig. 25.—Diagram Showing All Electrical System Parts of the Ford Model A Car and Their Location Relative to Each Other.

The Ford Model A Car

adjustment should be made with the engine warm. Do not expect a new engine that is stiff to "rock" on compression when stopped, or to idle perfectly at low speed.

Electrical System.—The electrical system of the Ford Model A includes the following equipment, as shown at Fig. 25: Storage Battery, Generator, Starting Motor, Distributor, Ignition Coil, Spark Plugs, Ammeter, Horn, Lamps, Windshield Wiper (on closed cars).

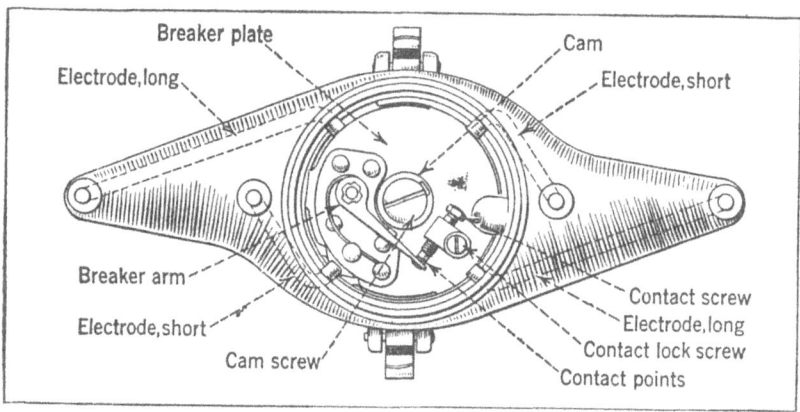

Fig. 26.—Top View of Ignition Current Distributor Showing Make and Break Points.

The current for igniting the gas mixture in the cylinders is provided by the storage battery. The ignition coil transforms the low tension current to a high tension current of sufficient voltage to jump between the points of the spark plugs. The distributor breaker points interrupt the flow of low tension current at regular intervals, while the distributor rotor distributes the high tension current to each spark plug in proper firing order. A complete wiring diagram is presented at Fig. 28.

Adjusting Breaker Contact Points.—The gap between

Model A Electrical System

the breaker points is set at .015 to .018 in. The gap should occasionally be checked to see that the points are properly adjusted. The points are shown at Fig. 26.

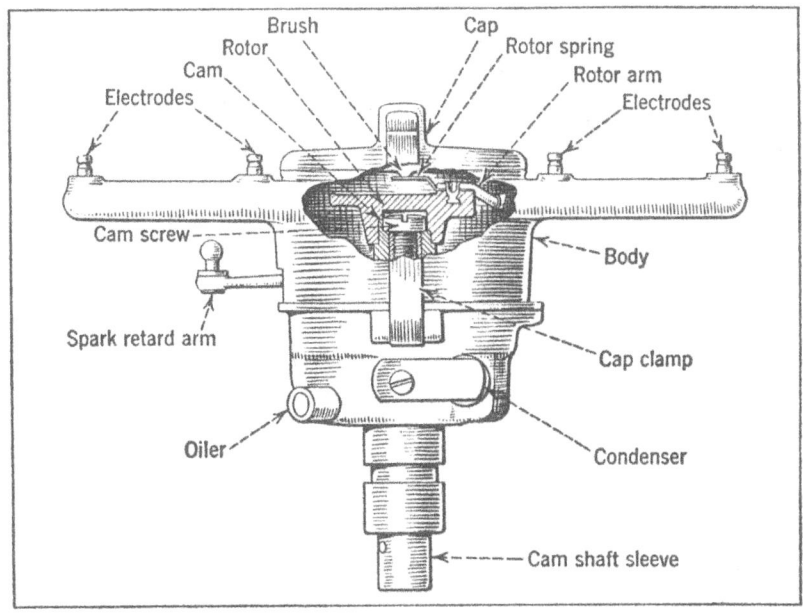

Fig. 27.—Side View of Ford Model A Ignition Distributor.

If the points are burnt or pitted, they should be dressed down with an oil stone. DO NOT USE A FILE.

To adjust the points, proceed as follows:

Lift distributor cap, rotor, and body.

Turn engine over slowly with starting crank until breaker arm rests on one of the four high points of the cam with the breaker points fully opened.

Loosen lock screw and turn the contact screw until the gap is at .015 to .018 in. A standard thickness gauge is used to obtain this measurement.

The Ford Model A Car

Ignition Timing.—As the spark must occur at the end of the compression stroke, the timing must be checked from that point. To find the compression stroke and time the spark, proceed as follows:

1. Fully retard spark lever.
2. Check gap between breaker contact points and, if necessary, adjust them as previously described.
3. Screw out timing pin located in timing gear cover and insert opposite end of pin into opening.
4. With the starting crank turn the engine over slowly, at the same time pressing in firmly on the timing pin. When the piston reaches the end of the compression stroke the timing pin will slip into a recess in the camshaft gear.
5. With the pin in place, remove the distributor cover and lift off rotor and distributor body.
6. Loosen cam-locking screw, shown at Fig. 27, until cam can be turned.
7. Replace rotor and turn it until the rotor arm is opposite No. 1 contact point in distributor head.
8. Withdraw rotor from cam and slightly turn the cam in a counter clockwise direction until the breaker points just start to open, then securely tighten cam-locking screw.
9. Replace rotor and distributor cover.
10. WITHDRAW TIMING PIN FROM RECESS IN TIME GEAR and screw it back tightly into the timing gear cover.

The Ford Starting System uses a six-volt, 80 ampere hour, 13-plate battery, designed and built in the Ford factory to meet the requirements of the Ford car.

Model A Storage Battery

Every two weeks check the electrolyte in the battery, to see that it is at the proper level. The solution (Electrolyte) should be maintained at a level with the bottom of the filling tube. If below this point, add distilled water until the electrolyte reaches the proper level. Water for battery use should be kept in clean, covered vessels of glass, china, rubber or lead. In cold weather add water only immediately before running the engine, so that the charging will mix the water and electrolyte and prevent freezing. Access to the battery is easily made by removing a small plate located in the floor board in front of the driver's seat. To remove the battery from the car it will be necessary to take out the floor boards. WHEN REPLACING THE BATTERY IN THE CAR BE SURE TO INSTALL IT WITH THE POSITIVE TERMINAL GROUNDED TO THE FRAME, AS SHOWN IN FIG. 25.

Keep the battery filling plugs and connections tight, and the top of the battery clean. Wiping the battery with a rag moistened with ammonia will counteract the effect of any of the solution which may be on the outside of the battery. A coating of vaseline will protect the terminals from corrosion. It is of vital importance that the battery is firmly secured in its supporting brackets at all times. If clamps are loose, the battery will shift about in the compartment and result in loose connections, broken cells, and other trouble. When repairs are necessary, or if the car is to be laid up for the winter, take the battery to a Ford dealer for proper attention and storage. Do not entrust your battery to inexperienced or unskilled hands.

The Generator.—The generator is of the Power House type, so called because of its similarity to the type of

The Ford Model A Car

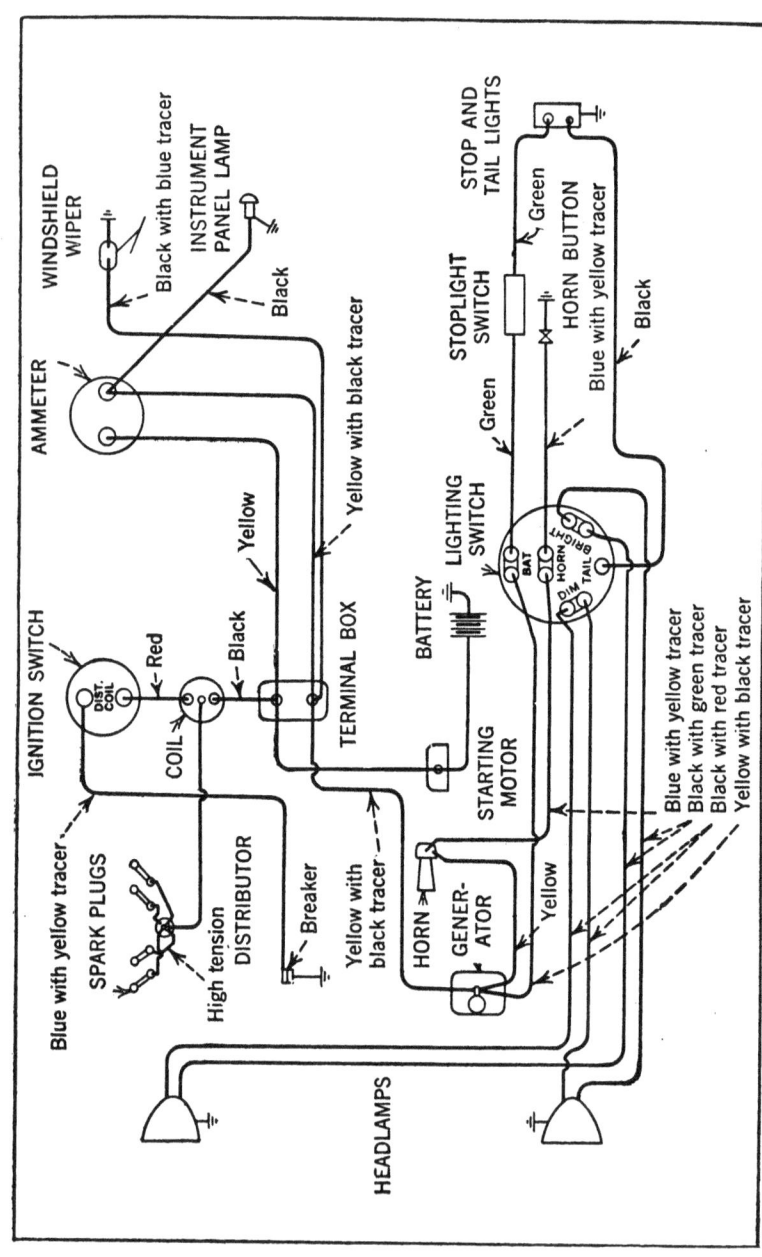

Fig. 28.—Wiring Diagram of the Ford Model A Showing All Units and Connections.

Model A Generator Adjustment

generator used in power houses. It is mounted on the left hand side of the engine. During winter months the charging rate should be adjusted to 14 amperes; in the summer this rate should be cut down to 10 amperes. The rate can, of course, be increased or decreased to meet individual requirements. For example, the owner who takes long daylight trips should cut down the charging rate to 8 amperes, to prevent the battery overcharging. On the other hand, the owner who makes numerous stops, should increase the normal rate if his battery runs down.

To increase or decrease the generator charging rate, remove generator cover and loosen field brush holder lock screw. The field brush holder can be easily identified, as it is the only one of the five brush holders that operates in a slot in the brush holder ring and which is provided with a locking screw. The remainder of the brush holders are riveted to the ring and are not movable. To increase the charging rate, shift the field brush holder in the direction of rotation; to reduce the rate shift the brush in the opposite direction. The output of the generator is indicated by the ammeter located on the instrument panel.

The starting motor is mounted on the left side of the engine. It requires no attention beyond seeing that the cable connection is clean and tight.

Ignition Coil.—The ignition coil mounted on the dash receives the low tension current from the battery, and transforms it into the high tension current necessary to produce the spark at spark plugs. Occasionally inspect the wire connections at the coil, distributor and spark plugs, to see that they are clean and tight. Loose connections cause misfiring.

The Ford Model A Car

The Spark Plugs.—The spark plugs are the medium through which the electric current ignites the gasoline charge in the cylinder. Hard starting or misfiring of the engine may be caused by dirty spark plugs or incorrect spark plug gaps. Keep the plugs clean and the gap set to .025 in. There is nothing to be gained by experimenting with different makes of spark plugs. The spark plugs with which Ford engines are equipped when they leave the factory are best adapted to the requirements of the Ford engine and have been determined to be so by exacting factory tests.

The Ammeter.—The ammeter is located on the instrument panel. It registers "charge" when the generator is charging the battery, and "discharge" when the lights are burning and the engine only running about 10 miles per hour. If the engine is running above 15 miles per hour and the ammeter does not register "charge," consult a Ford dealer or competent electrician specializing on automotive work.

Operation of the Lights.—The lighting system consists of a dash light, two head lamps and a combination tail and stop light, operated by a switch located at the top of the steering wheel. The headlamp bulbs are of the 6-8 volt, double filament gas-filled type. The major filament is 21 candle power. The small bulb used in the tail light, the instrument panel light, is of the 6-8 volt, single contact, three-candle power type. The stop lamp bulb is a single contact, 21 c.p. bulb. All of the lamps are connected in parallel so that the burning out or removal of any of them will not affect the others. Current for the lamps is supplied by the battery.

When replacing burned-out bulbs, make certain that you get genuine Ford bulbs, as satisfactory results can-

Model A Headlamp Alignment

not be obtained with the many inferior bulbs now on the market. Genuine Ford bulbs have the name FORD marked on the base. They insure your headlights meeting the lighting requirements of the various states.

Headlamp Requirements.—When the car is delivered, the headlamps are properly focused and aligned, and will pass the lighting requirements of all states. Should the lamps get out of focus or alignment, they should immediately be focused and realigned. Ford dealers are equipped to do this work, or if you have the proper facilities you can make the adjustment.

Align and focus headlamps with empty car standing on a level surface in front of a white wall or screen 25 feet from front of headlamps. This wall must be in semi-darkness or sufficiently shielded from direct light so that the light spots from the headlamps can be clearly seen. The wall must be marked off with black lines, as shown in Fig. 29 A and B. Details for making the layout are shown in Fig 29 C.

To Focus.—Turn on bright lights. Focus by means of screw at back of lamps; keep one lamp covered while focusing the other. Adjust the bulb filament at the focal center of the reflector to obtain an elongated elliptical spot of light on the wall, with its long axis horizontal. (See Fig. 29 A.) In focusing, adjust the bulb to obtain as good contrast and as well-defined cut-off across the top of the spot of light as possible. With lamps thus focused for the "bright" filament, the "dim" will be in satisfactory position.

Alignment.—Loosen nut at bottom of bracket and tilt headlamps to desired angle.

The tops of the bright spots on the 25-foot wall are to be set at a line 33 inches above level of surface on

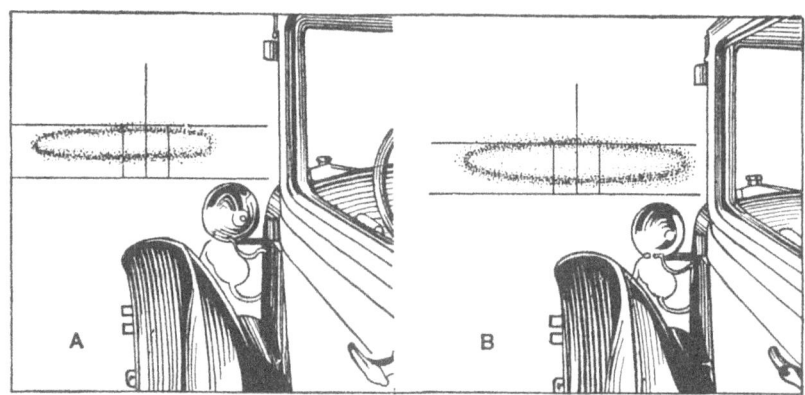

Left headlamp properly focused and aligned

Both headlamps properly focused and aligned

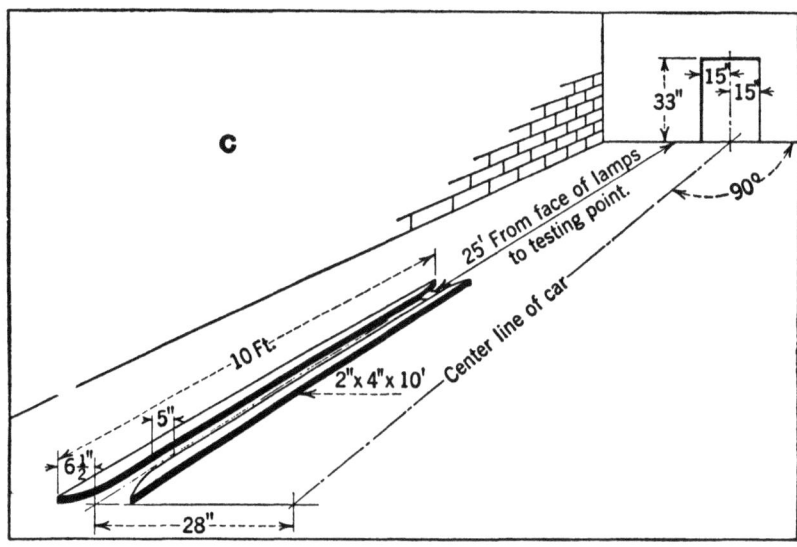

Fig. 29.—Methods of Focusing and Aligning Ford Model A Headlamps Shown at A and B. Shop Layout for Focusing and Adjusting Headlamps Shown at C.

Aligning Model A Headlamps

which car stands. With top line thus set for empty car, the headlamps will also have the proper tilt under full loads, as required by the various states.

The beam of light from each headlamp is to extend straight forward; that is, the centers of the elliptical spots of light must be 30 inches apart.

Proper alignment of headlamps is readily checked by means of a horizontal line on the wall in front of the car, 33 inches above the level surface on which the car stands, and two vertical lines 30 inches apart, each one 15 inches from center line of car (see Fig 29 C). Proper alignment of car relative to marks on the wall may be readily provided by use of wheel guide blocks for one side of the car, as shown in Fig 29 C. If it is impossible to tie up the floor space required by these blocks, marks painted on the floor may be used to show where one set of wheels should track and where the car should be stopped. In order to avoid any confusion, the new layout can be painted with red paint and the old that is needed with model T painted with black.

Care of Running Gear.—Every few weeks the front and rear axles should be carefully gone over to see that all nuts and connections are tight, with cotter pins in place. The spring clips should be inspected occasionally to see that they are tight.

The front wheels should be jacked up periodically and tested for smoothness of running and excessive side play. To determine if there is excessive side play, grasp the sides of the tire and shake the wheel. Do not mistake loose spindle bushings for loose bearings. Insert a cold chisel between spindle and axle when making this test to take up any spindle bushing play.

The Ford Model A Car

Adjusting Front Wheel Bearing.—If there is excessive play in the bearing, it can be adjusted as follows:

Remove wheel.

Withdraw cotter key and tighten adjusting nut until the hub just starts to bind.

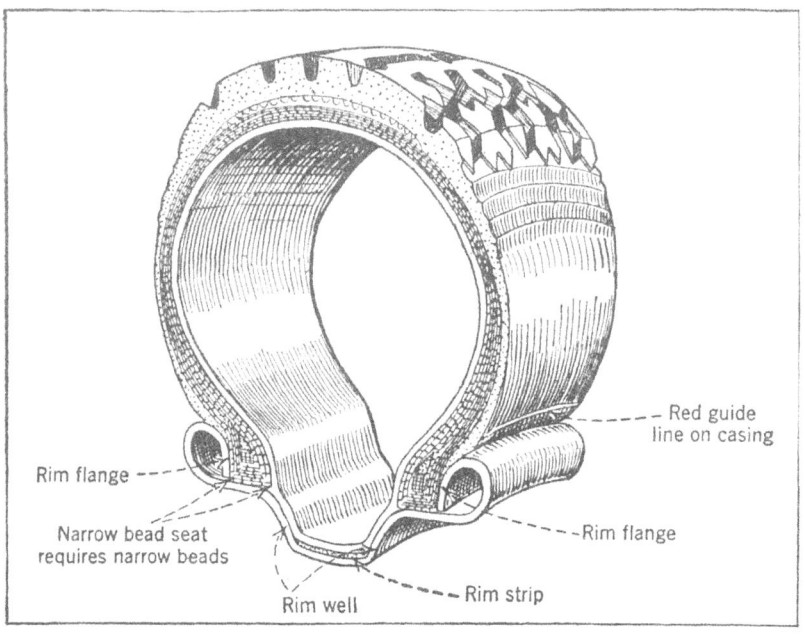

Fig. 30.—Sectional View Showing Drop Center Rim Used with Ford Wire Wheels and Balloon Tire Installed Thereon.

Then back off the adjusting nut one or two notches until the hub can be freely revolved.

Before placing the wheel, BE SURE TO INSERT COTTER KEY IN ADJUSTING NUT.

The springs should be lubricated occasionally with oil or graphite. This will restore the original flexibility of the springs and improve the riding quality of the car.

Adjusting Model A Front Wheel

Ford Wire Wheels.—To remove the Ford wire wheels, jack up the side of the car from which the wheel is to be withdrawn and screw off the five hub bolt nuts. The wheel can then be removed. When replacing a wheel, tighten each hub bolt nut a few turns at a time. Then follow around hub, tightening each nut firmly. If nuts are not drawn up evenly, the wheel will not run true.

Removing Tires From Ford Wire Wheels (Drop Center Rims).—Remove valve cap and lock nut and place wheel so that the valve is at the top. Let all air out of tube. Push valve stem up into tire. Working both ways from the valve stem, press the casing together and down into the rim well, approximately one foot each side of the valve stem. Insert tire iron under both beads at point opposite valve and force tire over rim. The tire can be then removed from the wheel with the hands.

Mounting Tire on Ford Wire Wheels (Drop Center Rims).—Inflate tube until it is barely rounded out, and insert tube in casing. (CAUTION: Never use a tire flap when mounting tires on a Ford wire wheel.) With wheel placed so that valve stem hole is at top, place casing and tube on wheel, with valve in valve stem hole. (See Fig. 31 A.) Working both ways from the valve stem, press the casing together and down into the rim well, until lower part of casing can be forced over rim flange at bottom. A tire iron may be used, if necessary. (See Fig 31 B.) Raise tire up (see Fig. 31 C) until it is perfectly centered on rim and beads are seated on bead seats. Inflate tube to not more than two pounds pressure and work casing back and forth to insure proper setting of tire, indicated by red line on tire being equally spaced from rim all way around. (See Fig 31 D.) CAUTION: With Ford wire wheels (drop center rims) use only casing

The Ford Model A Car

with red centering line just above the rim flange, and tubes marked "for drop center rims."

It is particularly important that the red line (Fig.

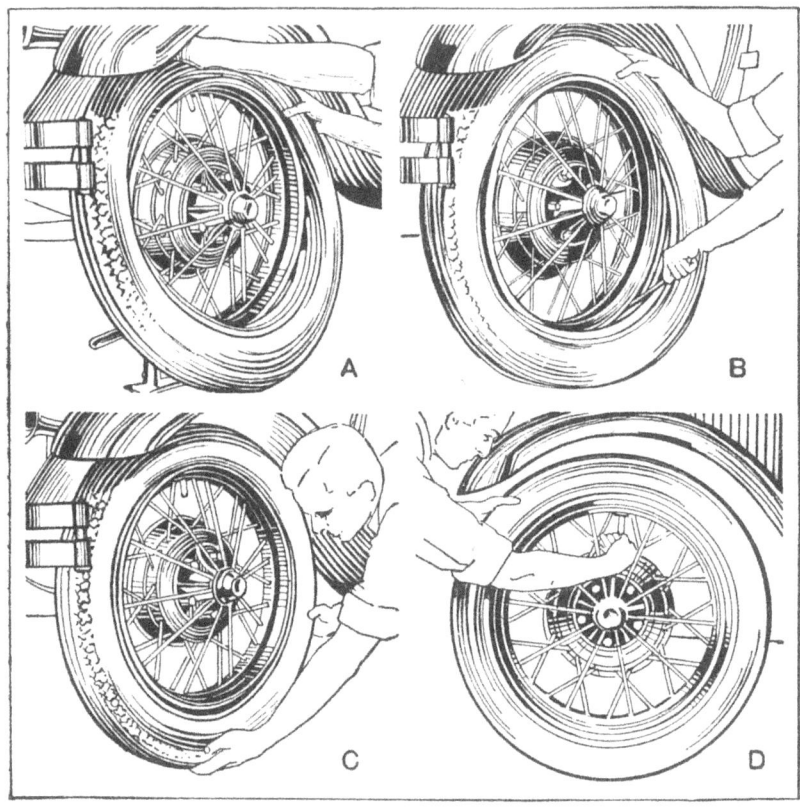

Fig. 31.—Sketches Showing Manipulation of Tire on Ford Drop Center Rim.

30) shows an even distance from the rim all around on both sides before fully inflating tire. Put valve nut on valve, inflate tire to 35 pounds and screw valve cap down tightly. (See Fig. 31 D.) With Ford wire wheels, tires

Model A Tire Inflation and Care

can be more easily changed with wheel mounted on axle or tire carrier than by laying the wheel down on ground.

Tires should never be run partially inflated, as the side walls are unduly bent and the fabric is subjected to stresses which cause what is known as rim cutting. KEEP BOTH FRONT AND REAR TIRES INFLATED TO 35 POUNDS AND CHECK THE PRESSURE ONCE A WEEK. Never run on a flat tire, even for a short distance. Skidding also shortens the life of the tires. Avoid locking the wheels with the brakes—no tire will stand the strain of being dragged over pavement. Avoid running in street car tracks, or bumping the sides of the tire against the curbing. To get most service at least expense, tires should be inspected frequently and all small cuts or holes properly sealed or repaired, thus preventing dirt and water working in between the rubber tread and the fabric, causing blisters or sand holes. The construction of Ford tire and drop side rim is clearly shown at Fig. 30.

When a car is idle for any appreciable length of time, it should be jacked up to take the load off the tires. If the car is laid up for several months, it is best to remove the tires. Wrap up the outer casings and inner tubes separately, and store them in a dark room, not exposed to extreme temperatures. Remove oil or grease from the tires with gasoline.

Brakes.—The braking system consists of four internal expanding brakes—one brake located within each wheel. The brakes are brought into operation by depressing the foot brake pedal or pulling back the hand brake lever. Each brake consists of two rigid shoes faced with lining; a hardened steel brake drum and special operating and adjusting parts. (See Fig. 32.) When the brakes are

The Ford Model A Car

applied the brake-shoes are expanded against the drum, the pressure of the shoes setting up a strong braking resistance which stops the car. An operating cam expand the shoes in the rear wheels, while in the front wheels a wedge is used.

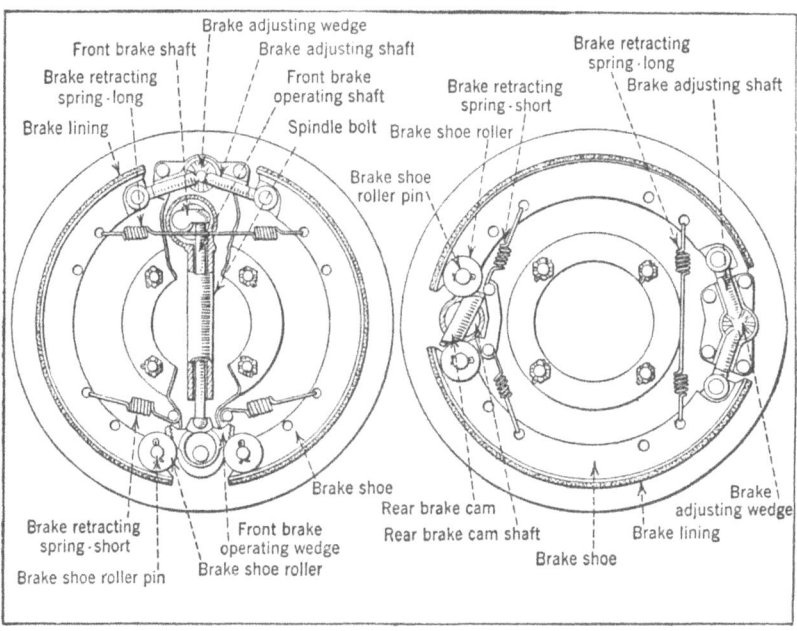

Fig. 32.—Parts of Ford Model A Internal Expanding Wheel Brakes and Shoe Spreading Mechanism.

Ordinary adjustments necessitated by natural wear of the linings are made as follows:

Jack up one wheel at a time and with the hand brake lever fully released turn the adjusting wedge (see Fig. 33) on each wheel until the wheel starts to bind. Then back off the adjusting wedge one notch, or until this wheel revolves freely. Each wheel should be adjusted in this

Adjusting Model A Brakes

manner. When all of the adjustment on the adjusting wedges is used up, it will be necessary to reline the brakes. When this becomes necessary we suggest you take your car to an authorized Ford dealer. They are provided with special relining equipment.

UNDER NO CIRCUMSTANCES SHOULD OWN-

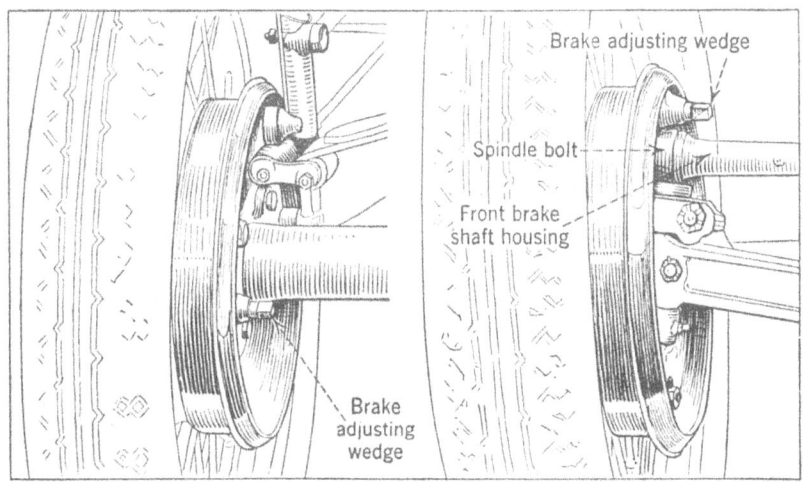

Fig. 33.—Where the Ford Model A Brake Adjusting Wedges Are Located.

ERS ATTEMPT TO MAKE ADJUSTMENTS BY TURNING UP THE CLEVISES ON THE BRAKE RODS, AS THIS WOULD RESULT IN THE BRAKES BEING THROWN OUT OF ADJUSTMENT AND CAUSE UNEQUAL WEAR ON THE LININGS.

Shock Absorbers.—Ford hydraulic double-acting shock absorbers operate entirely on the principle of hydraulic resistance, and are shown installed at Fig. 34. Glycerine is forced from one chamber to another by the movement of the lever arm. The working chamber is automatically kept full by the glycerine in the reservoir. As the shock

The Ford Model A Car

absorbers are accurately adjusted at the factory, it should not be necessary to alter this adjustment except in rare cases, where more or less shock absorber action is desired.

The reader will observe a needle valve with an arrow pointer extending through the center of the shaft. Surrounding it numbers from 1 to 8 are stamped. The average setting is with arrow pointing at 2 for the front

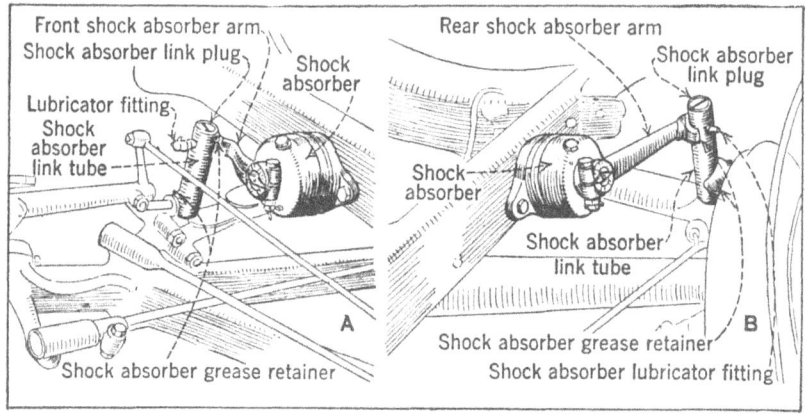

Fig. 34.—Location and Parts of Ford Model A Hydraulic Shock Absorbers.

shock absorbers and 3 for the rears. Turning the needle valve changes the adjustment. Resistance is increased when the needle valve is turned from 1 to 8 and decreased from 8 back to 1. A slight movement of the needle valve either way makes a big difference in the action of the instruments.

The only care the shock absorbers require is replenishing the glycerine in the reservoir and the lubrication of the ball joints. The filler plug in the reservoir should be removed at intervals of 5,000 to 10,000 miles, and the reservoir filled with glycerine (Commercial). NEVER

Model A Shock Absorbers

REPLENISH WITH OIL. Oil will solidify in the winter or reduce resistance, and will not mix with the glycerine in the instruments. In warm climates, replenish with glycerine, C. P. or Commercial. All instruments contain glycerine with 10% alcohol. Where temperatures of zero and below are prevalent, add additional alcohol, not to exceed 20% of the volume.

The ball joint in itself is made in unit with the instrument arm. It is hardened and ground. The ball joint seats are enclosed in the shock absorber connecting links, which should be lubricated every 500 miles with the compressor gun. In order to secure maximum riding comfort, it is important that the spring hangers be free in the bushings and kept well greased.

The speedometer with which the Ford Model A car is equipped indicates the speed and records the mileage of the car. It also helps the user in maintaining an accurate lubrication schedule. To reset the trip odometer, simply pull out the knob on the speedometer. By turning the knob you can reset the figures at any tenth of a mile desired, or back to zero. After resetting the odometer, be sure to push the knob back to its original position.

The flexible shaft should be lubricated every 5,000 miles. Care should be exercised not to bend the shaft in a radius smaller than 7 in. For maximum service we recommend that every 10,000 miles the speedometer be cleaned, lubricated and recalibrated. When this is necessary, or when repairs are required, consult an authorized dealer. All speedometers are sealed when they leave the factory. Under no circumstances should this seal be broken or repairs attempted by the owner.

The Proper Way to Wash the Car.—Always use cold or lukewarm water—never hot water. If a hose is used, don't turn the water on at full force. This drives the dirt into the finish. After the surplus mud and grime have been washed off, take a sponge and clean the body and running gear with a solution of water and linseed oil soap. Rinse off with cold water; then rub and polish the body with a damp chamois skin. A body polish of good quality may be used to add luster to the car. Grease on the running gear may be removed with a gasoline-soaked sponge or rag. The nickeled parts may be polished with a good nickel polish. An excellent body polish and nickel polish can be purchased from any Ford dealer.

Should the body or other pyroxylin finished parts of the car become spattered with tar or other substances used on roads, the spots can be removed with a solution of two-thirds gasoline and one-third oil. If full strength gasoline is used, there is a possibility of softening the finish. Dip a soft cloth into the mixture and, using one finger, rub the spot gently until it has been removed. The rubbed spot should then be washed off with clear water.

Care of the Top.—When putting down the top be careful in folding to see that the fabric is not pinched between the bow spacers, as they will chafe a hole through the top very quickly. Applying Ford top dressing will greatly improve the appearance of an old top.

Storing the Car.—Drain the water from the radiator, then put in about a quart of anti-freeze solution to prevent any freezing of the water that may possibly remain. Draw off all gasoline. Drain the old oil from the oil pan. Refill the oil pan with one gallon of fresh oil and run the engine enough to cover the different parts with oil. Remove the tires and store them. Wash the car and, if possible, cover the body with a sheet of muslin to protect the finish.

Care of Model A Truck

FORD MODEL A ONE AND ONE-HALF TON TRUCK

The care and general operating instructions pertaining to the car also apply to the truck, with the exception that a new truck should not be driven faster than 20 to 25 miles per hour for the first 500 miles, and all tires

Fig. 35.—Location of Dual High Transmission Control Pedals on Ford Model AA Truck.

should be inflated to 80 pounds and the pressure checked every week.

Dual High.—The dual high transmission, which is optional equipment, gives the truck approximately one-third more pulling power. It is operated by a double end shift pedal which extends through the floor boards. Pressing down on the rear pedal engages the dual high. Pressing down on the front pedal disengages it. (See Fig. 35.)

The Ford Model A Car

This shift is entirely independent of any standard transmission gear shift and can be made at any speed in which the truck is being driven, whether high, second, low or reverse.

To engage the dual high remove your foot from the accelerator. Disengage the clutch, then press down on the rear dual high clutch pedal. As soon as the dual high is engaged, re-engage the clutch and press down on the accelerator until the desired driving speed is obtained.

To disengage the dual high, remove your foot from the accelerator. Disengage the clutch and press down on the front dual high clutch pedal. As soon as the dual high is disengaged, re-engage the clutch and press down on the accelerator until the desired driving speed is obtained.

SUMMARY OF FORD MODEL A ENGINE TROUBLES AND THEIR CAUSES

If starter turns engine over freely, check the following:

Ignition switch off.

Gasoline tank is empty or supply shut off.

If engine is cold, mixture may not be rich enough—choke button not pulled back. See starting instructions.

Warm engine—over-choking.

Breaker points too close. The correct adjustment is .015 to .018 in.

Spark plugs gaps too wide. Correct gap .025 in.

Water in sediment bulb or carburetor. See instructions for draining carburetor.

Starter Fails to Turn Engine Over.—Battery run down. A quick way to check this is to turn on the lights, and

Summary of Model A Engine Faults

depress the starter switch. If the battery is weak the lights will go out or grow quite dim. If the battery is run down, have it recharged.

Loose or dirty battery connections—see that both the negative and positive battery terminal connections are clean and tight. These connections should be checked regularly.

Missing at Low Speed. —Gas mixture too rich or too lean. See carburetor adjustment on page 547.

Too close a gap between spark plug points. The correct gap is .025 in.

Breaker points improperly adjusted, badly burnt or pitted. See instructions for adjusting breaker contact points. Fouled spark plug. Plugs should be occasionally cleaned and the gaps checked. Water in gasoline. See instructions on cleaning sediment bulb and carburetor.

Missing at High Speed. — Insufficient gasoline flowing to carburetor due due to gasoline line filter screen being partly clogged. Gas mixture too rich or too lean. See carburetor adjustment instructions. Water in gasoline. Drain sediment bulb and carburetor as described.

Engine Stops Suddenly.—Gasoline tank empty. Dirt in fuel line or carburetor. Gas mixture too lean.

Engine Overheats. —Lack of water—radiator should be kept well filled. Lack of oil—check oil level as described. Fan belt loose or slipping. See instructions for fan belt adjustment. Carbon deposit on piston heads and in combustion chamber. This can be corrected by taking off the cylinder head and removing the carbon. (Ford dealers are equipped for this work.) Incorrect spark timing. Follow ignition timing instructions. Gas mixture too rich. See instructions for adjustment of carburetor. Water circulation retarded by sediment in radiator.

The Ford Model A Car

Engine Knocks.—Carbon knock—caused by a deposit of carbon in combustion chamber and on piston heads. Take off cylinder head and remove carbon. Ignition knock—if this occurs under ordinary driving conditions, check ignition timing. Engine overheats. Check conditions listed under "Engine Overheats." Loose bearing. If a bearing has become loose, it should be adjusted by an authorized Ford mechanic.

Do not mistake an ignition knock for a loose bearing. Ignition knocks usually occur when the car is suddenly accelerated or when ascending a steep grade or traveling through heavy sand with the spark lever fully advanced. Slightly retarding the spark lever eliminates the knock. The spark should be advanced as soon as normal road conditions are encountered. Loose bearings knock all the time the engine is in use and regardless of spark timing.

NOTES

NOTES

NOTES

NOTES

NOTES

www.ingramcontent.com/pod-product-compliance
Lightning Source LLC
Chambersburg PA
CBHW071702090426
42738CB00009B/1627